T0133883

Computational Biology

A Statistical Mechanics Perspective

Second Edition

Chapman & Hall/CRC Mathematical and Computational Biology

About the Series

This series aims to capture new developments and summarize what is known over the entire spectrum of mathematical and computational biology and medicine. It seeks to encourage the integration of mathematical, statistical, and computational methods into biology by publishing a broad range of textbooks, reference works, and handbooks. The titles included in the series are meant to appeal to students, researchers, and professionals in the mathematical, statistical, and computational sciences and fundamental biology and bioengineering, as well as interdisciplinary researchers involved in the field. The inclusion of concrete examples and applications and programming techniques and examples is highly encouraged.

Series Editors

Xihong Lin
Mona Singh
N. F. Britton
Anna Tramontano
Maria Victoria Schneider
Nicola Mulder

Gene Expression Studies using Affymetrix Microarrays
Hinrich Gohlmann, Willem Talloen

Big Data in Omics and Imaging
Associated Analysis
Momiao Xiong

Introduction to Proteins
Structure, Function, and Motion, Second Edition
Amit Kessel, Nir Ben-Tal

Big Data in Omics and Imaging
Integrated Analysis and Casual Inference
Momiao Xiong

Computational Blood Cell Mechanics
Road Towards Models and Biomedical Applications
Ivan Cimrak, Iveta Jancigova

Stochastic Modelling for Systems Biology, Third Edition
Darren J. Wilkinson

An Introduction to Systems Biology
Design Principles of Biological Circuits, Second Edition
Uri Alon

Computational Biology
A Statistical Mechanics Perspective, Second Edition
Ralf Blossey

For more information about this series please visit: https://www.crcpress.com/Chapman--HallCRC-Mathematical-and-Computational-Biology/book-series/CHMTHCOMBIO

Computational Biology

A Statistical Mechanics Perspective

Second Edition

Ralf Blossey

CRC Press
Taylor & Francis Group
Boca Raton London New York

CRC Press is an imprint of the
Taylor & Francis Group, an **informa** business

CRC Press
Taylor & Francis Group
6000 Broken Sound Parkway NW, Suite 300
Boca Raton, FL 33487-2742

© 2019 by Taylor & Francis Group, LLC
CRC Press is an imprint of Taylor & Francis Group, an Informa business

No claim to original U.S. Government works

Printed on acid-free paper

International Standard Book Number-13: 978-1-138-58786-1 (Hardback)

This book contains information obtained from authentic and highly regarded sources. Reasonable efforts have been made to publish reliable data and information, but the author and publisher cannot assume responsibility for the validity of all materials or the consequences of their use. The authors and publishers have attempted to trace the copyright holders of all material reproduced in this publication and apologize to copyright holders if permission to publish in this form has not been obtained. If any copyright material has not been acknowledged please write and let us know so we may rectify in any future reprint.

Except as permitted under U.S. Copyright Law, no part of this book may be reprinted, reproduced, transmitted, or utilized in any form by any electronic, mechanical, or other means, now known or hereafter invented, including photocopying, microfilming, and recording, or in any information storage or retrieval system, without written permission from the publishers.

For permission to photocopy or use material electronically from this work, please access www.copyright.com (http://www.copyright.com/) or contact the Copyright Clearance Center, Inc. (CCC), 222 Rosewood Drive, Danvers, MA 01923, 978-750-8400. CCC is a not-for-profit organization that provides licenses and registration for a variety of users. For organizations that have been granted a photocopy license by the CCC, a separate system of payment has been arranged.

Trademark Notice: Product or corporate names may be trademarks or registered trademarks, and are used only for identification and explanation without intent to infringe.

Library of Congress Cataloging-in-Publication Data

Names: Blossey, Ralf, author.
Title: Computational biology : a statistical mechanics perspective /
Ralf Blossey.
Description: Second edition. | Boca Raton, Florida : CRC Press, [2020] |
Series: Chapman & Hall/CRC mathematical and computational biology |
Includes bibliographical references and index.
Identifiers: LCCN 2019011496| ISBN 9781138587861 (hbk : alk. paper) |
ISBN 9780429503672 (ebk)
Subjects: LCSH: Biomolecules--Mechanical properties. | Statistical
mechanics. | Computational biology.
Classification: LCC QH506 .B57 2020 | DDC 572.80285--dc23
LC record available at https://lccn.loc.gov/2019011496

Visit the Taylor & Francis Web site at
http://www.taylorandfrancis.com

and the CRC Press Web site at
http://www.crcpress.com

To SyMaNi and Mum

Contents

Preface to the Second Edition

I am grateful to the publisher who asked me to renew my book as a second edition. Indeed, after more than ten years, the field of computational biology/statistical physics—the particular interface of two fields I am interested in—has matured in many ways. Going through the first edition again was a great pleasure when comparing past views to current insights.

As a consequence, the book has substantially changed while maintaining its core contents. New material has been added, but in particular I have reworked its whole architecture. In the second edition, there are only two parts: one on equilibrium statistical physics and one on non-equilibrium statistical physics. The theoretical tools in these fields that the book introduces are placed at the beginning of each part, and then the applications follow. I hope that this new structure will help the readers to master the material.

As a side effect of following the now classic dichotomy of statistical physics—equilibrium and non-equilibrium—the focus on *universal aspects*, those that statistical physics is most attuned to, has been sharpened. Even more than before, this makes the book *not* a biophysics book. Since 2006, when the first edition went into print, new books on biophysics have appeared, especially bioinformatics books.

I hope that with its modifications, including updated additional notes to the literature and references, this book will continue to find its readership and help to introduce its readers to this exciting field of research.

Lille, February 2019
Ralf Blossey

Preface to the First Edition

This is not a biophysics book.

The readers of this book will indeed find a number of topics in it which commonly classify as biophysics; an example is the discussion of the electrostatic properties of biomolecules. But the book's ambition is different from being yet another introduction into biophysics. I therefore like to explain my motivation for the selection of topics I made.

The title of the book establishes a link between computational biology on the one side, and statistical mechanics on the other. Computational biology is the name of a new discipline. It is probably fair to say that it is, in fact, an emerging one, since a new name is not sufficient to establish a new discipline. Quantitative methods have been applied to biological problems since many years; this has led to a vast number of different subdisciplines of established fields of science: there is a mathematical biology, a biomathematics, a biostatistics, a bioinformatics, a biophysics, a theoretical biology, a quantitative biology... this list is certainly not exhaustive.

All these subdisciplines emerged out of the existing fields and developed by an act of transfer: the use of a method or a mathematical approach within a new context, its application to new problems. One may expect that at some point in the future all these different subdisciplines may merge into a common discipline. For the time being and for the lack of a definite name, let us call this discipline computational biology.

This book wants to contribute a particular element to this field; the use of statistical mechanics methods for the modelling of the properties of biological systems. Statistical physics is the scientific discipline which was developed in order to understand the properties of matter composed of many particles. Traditionally, it has been applied to non-living matter: gases, liquids, and solids. Meanwhile, it is increasingly applied to what is nowadays called "soft matter" which encompasses complex objects like colloids, membranes, and biomolecules, hence objects which do not clearly fall into any one of the classic categories.

Statistical mechanics methods are therefore indeed essential in their application to biophysical problems, since they are needed to understand the static and dynamic properties of biomolecules, complex molecular machines, and even whole cell behaviour.

But there is a second aspect for which these methods can prove important, and this relates to the information content of the biological systems. Biology is built on recognition processes: DNA strands have to recognize each other, proteins have to identify DNA binding sites, etc. In bioinformatics, these recognition problems are commonly modelled as pattern recognition problems: this mapping is the basis of the enormous success of the field of modern genomics.

Moving beyond genomics, however, to the biology of whole systems, it becomes increasingly clear that an understanding of the physical properties of biological systems becomes more and more important for processes involving biological information content: DNA is not merely a string spelled out in a four-letter alphabet, but it is also an elastic string. There is biological information contained in its *structure* and its *dynamics*. Biology employs physical mechanisms to organize its information processes. This becomes particularly evident, as we begin to understand, in the properties of chromatin, the DNA-protein complex making up the chromosomes in the nucleus of eukaryotic cells (a topic which is discussed in the book), or, in the particle interaction networks upon which the cellular machinery relies.

This book is placed at just this interface: between biological recognition on the one hand, and the statistical physics methods that can be employed to understand the underlying mechanisms.

The first part of the book gives a concise introduction into the main concepts of statistical mechanics, equilibrium, and non-equilibrium. The exposition is introductory in the choice of topics addressed, but still mathematically challenging. Whenever possible, I have tried to illustrate the methods with simple examples.

The second part of the book is devoted to biomolecules, to DNA, RNA, proteins, and chromatin, i.e., the progression of topics follows more or less what is commonly known as the central dogma of molecular biology. In this part, mostly equilibrium statistical mechanics is needed. The concern here is to understand and model the processes of base-pair recognition and (supra-) molecular structure formation.

The third part of the book is devoted to biological networks. Here, both equilibrium and non-equilibrium concepts introduced in the first part are used. The presentation covers several of the non-equilibrium statistical physics approaches described in Chapter 2 of Part I, and illustrates them on biologically motivated and relevant model systems.

Throughout the book, *Exercises* and *Tasks* are scattered, most of them in the first part. They are intended to motivate the readers to participate actively in the topics of the book. The distinction between *Exercises* and *Tasks* is the following: *Exercises* should be done in order to verify that the concept that was introduced has been understood. *Tasks* are more ambitious and usually require either a more involved calculation or an additional idea to obtain the answer.

A final technical note: the book is complemented by a detailed key word list. Key words listed are marked in italics throughout the text.

In the course of shaping the idea of this book and writing it, I profited from discussions with many colleagues. I am especially thankful to Arndt Benecke, Dennis Bray, Martin Brinkmann, Luca Cardelli, Enrico Carlon, Avi Halperin, Andreas Hildebrandt, Martin Howard, Oliver Kohlbacher, Hans Meinhardt, Thomas Lengauer, Hans-Peter Lenhof, Annick Lesne, Ralf Metzler, Johan Paulsson, Andrew Phillips, Wilson Poon, Helmut Schiessel, Bernard Vandenbunder, Jean-Marc Victor, Pieter Rein ten Wolde, Edouard Yeramian.

Finally, I would particularly like to thank Andreas Hildebrandt and Martin Howard for their detailed and helpful remarks on an early version of the manuscript.

I gratefully acknowledge the hospitality of the Institut d'Électronique, de Microélectronique et de Nanotechnologie (IEMN) in Villeneuve d'Ascq and the Institut des Hautes Études in Bures-sur-Yvette, where parts of this book were written.

Lille, 2006

I

Equilibrium Statistical Mechanics

CHAPTER 1

Equilibrium Statistical Mechanics

Après tout, vous êtes ici, c'est l'essentiel! Nous allons faire du bon travail ensemble [...]! D'ici peu, le monde ébloui decouvrira la puissance du grand Z!

Franquin, Z comme Zorglub (1961)

1.1 Z: THE PARTITION FUNCTION

This section introduces the basic physical concepts and mathematical quantities needed to describe systems composed of *many particles*, when only a statistical description remains possible.

Already for an apparently trivial system such as a gas of identical atoms, composed of $N \sim 10^{23}$ particles/mole (*Avogadro's number*), the description of each particle's trajectory is illusory and not even desirable, and hence only a statistical approach feasible. In biology, as we will see, things are more complicated: not only is the number of relevant interacting components large, but in addition the components are very often 'individuals'.

Suppose we can characterize our system of interest by a given number of 'state variables', which can be finite or infinite. An example of such a state variable can be the spatial position of a particle, hence a continuous variable, but in principle it can be any other of its distinguishing characteristics. We call the set of these state variables **x**, irrespective of their physical nature, and treat **x** here as a set of discrete variables;[1] we call such a collection of variables characterizing the state of a system a *microstate*. The possible microstates a system can assume, typically under certain constraints, can be subsumed under the name of a *statistical ensemble*. One has to distinguish the *microstate* the system can assume in each of its realisations from the ultimate *macrostate* that is to be characterized by a macroscopic physical observable after a suitable averaging procedure over the microstates.[2]

Example. We give a simple illustration of the concept by considering the conformations of a linear molecule such as DNA. The macrostate is the chain conformation we will most probably observe when we image the molecule under certain experimental conditions, while a microstate is any mechanically possible conformation of the molecule. The collection of all these possible states is what we call the statistical ensemble.

Coming back to the conceptual question: what quantity governs the probability $P(\mathbf{x})$ of each state to occur?

Probability and information. We call $P(\mathbf{x}) \geq 0$ a *probability*; consequently, $\sum_{\{\mathbf{x}\}} P(\mathbf{x}) = 1$, where the summation is over all possible states. To each probability we can associate an *information measure*[3]

$$s(\mathbf{x}) = -\ln P. \tag{1.1}$$

For $P = 1$, $s = 0$, and for $P = 0$, $s = \infty$. Therefore $s(\mathbf{x})$ is a measure for the *lack of information*. It can be used to define *entropy* as the average of $s(\mathbf{x})$ in distribution,

$$S(P) = \sum_{\{\mathbf{x}\}} s(\mathbf{x})P(\mathbf{x}) = -\sum_{\{\mathbf{x}\}} P(\mathbf{x})\ln P(\mathbf{x}). \tag{1.2}$$

In our terminology entropy can also be called an *average ignorance*. The information-theoretic unit of entropy is the *bit*.

---

[1] In the case of continuous variables, sums have to replaced by integrals in an obvious way; we will encounter this later and will pass liberally from one notation to the other.

[2] The averaging procedure means that we can determine the macroscopic state by an average over the ensemble of microstates; the resulting *ensemble average* gives a correct description of the macrostate of the system if the system had sufficient time in its dynamic evolution to sample all its microstates; this condition is called *ergodicity*.

[3] The choice for the logarithm will become clear in the section on thermodynamics. Note that we do not distinguish in our notation between log and ln, the meaning should be clear from the context.

Example. Again we take as a simple example DNA, which consists of the four bases A, C, G, T (the details of the build-up of DNA are explained in Chapter 2). If the probability of all bases is identical, $P(base) = 1/4$, and we have $S(P) = 4 \times (1/4) \times \ln_2(4) = 2$. The average ignorance is thus 2 bits per nucleotide. In order to interpret this result consider the following: before a base is read by some 'device' (e.g., a polymerase transcribing DNA into RNA), the information can be represented in binary code as two bits: 11, 10, 01, 00. After reading, the uncertainty has become 0.

Maximum entropy. What determines the form of P? In order to deduce it, we employ the prescription proposed by E. T. JAYNES, 1957. It states that the *prior probability distribution* maximizes entropy while respecting macroscopic constraints; it should thus yield the *maximum average ignorance* under those constraints.

Consider first the case of equiprobabilities, as we have assumed for our simple DNA example: there is no other constraint. In order to apply Jaynes' concept, we have to maximize $S(P)$ under the only natural constraint of normalization, $\sum_{\{\mathbf{x}\}} P(\mathbf{x}) = 1$. This leads to a variational problem[4]

$$\delta[S + \lambda[\sum_{\{\mathbf{x}\}} P(\mathbf{x}) - 1]] = 0 \tag{1.3}$$

in which δ denotes the *variation* of the bracketed term with respect to P; e.g., one has $\delta S(P) = -\sum_{\{\mathbf{x}\}} [\delta P \cdot \ln P + P \cdot \delta(\ln P)]$, and $\delta \ln P = [d(\ln P)/dP]\delta P$. In Eq. (1.3), λ is a *Lagrange multiplier*, a constant to be determined in the calculation.

This variational problem Eq. (1.3) is solved by

$$P(\mathbf{x}) = e^{\lambda - 1} = const. \equiv \Omega^{-1} \tag{1.4}$$

where Ω is the number of realizations of $\mathbf{x}$. All states are indeed found to be equally probable, with a probability inverse to their number. From the definition of entropy, Eq. (1.2), we have the relation

$$S = \ln \Omega. \tag{1.5}$$

Exercise. Show that $P(\mathbf{x})$ solves (1.3).

---

[4]For those confused by the δ-notation common to variational calculations, the same result is obtained by replacing $P \rightarrow P_0 + \epsilon \times P$ with a small parameter ϵ, and requiring the terms of linear order in ϵ to vanish.

The canonical ensemble. We now repeat the above argument and calculation for the so-called *canonical ensemble*. Here, one additional constraint appears, since we want to characterize the states of the system now by an *energy* $E(\mathbf{x})$, which we allow to fluctuate with average $\langle E \rangle = E_0$.

In this case we must maximize $S(P)$ under two constraints, the normalization condition, and in addition the condition on the average energy

$$\langle E \rangle = \sum_{\{\mathbf{x}\}} E(\mathbf{x}) P(\mathbf{x}) . \tag{1.6}$$

The resulting variational problem involves two Lagrange parameters, λ and β:

$$\delta[S + \lambda[1 - \sum_{\{\mathbf{x}\}} P(\mathbf{x})] + \beta[\langle E \rangle - \sum_{\{\mathbf{x}\}} E(\mathbf{x}) P(\mathbf{x})]] =$$
$$= \sum_{\{\mathbf{x}\}} \delta P(\mathbf{x}) \cdot (-\ln P - 1 - \lambda - \beta E(\mathbf{x})) = 0 , \tag{1.7}$$

with the *canonical* or *Gibbs distribution* as a result,

$$P_\beta(\mathbf{x}) = Z_\beta^{-1} e^{-\beta E(\mathbf{x})} , \tag{1.8}$$

where Z_β is the *partition function*

$$Z_\beta = \sum_{\{\mathbf{x}\}} e^{-\beta E(\mathbf{x})} . \tag{1.9}$$

For dimensional reasons, the Lagrange parameter β must be an inverse energy, which we take as the *thermal energy*

$$\beta^{-1} = k_B T \tag{1.10}$$

with Boltzmann's constant k_B. We define the *free energy* as

$$F \equiv -k_B T \ln Z . \tag{1.11}$$

The meaning of this definition will become clear in the following section. Eq. (1.9) is the key quantity in all of the first chapter of this part of the book; we will be mostly concerned with methods how to compute it.

Task. Determine $P_{\beta,\mu}$ if not only the average energy $\langle E \rangle$ is taken as a constraint condition, but also the number of particles N is allowed to fluctuate with average $\langle N \rangle$. The associated Lagrange parameter μ is called *chemical potential.*[5] The ensemble governed by the resulting distribution is called the *grand canonical ensemble.*

Equivalence of ensembles. The different ensembles (canonical, grand canonical etc.) are macroscopically equivalent for equilibrium states in the *thermodynamic limit* of N, $V \to \infty$ with $n = N/V = const$. Deviations from the thermodynamic limit scale as $\sim 1/\sqrt{N}$, and are hence negligible since the value of N we usually talk about in statistical mechanics is on the order of Avogadro's constant.[6] This equivalence allows to choose the ensemble based on its convenience in a particular application.

1.2 RELATION TO THERMODYNAMICS

The knowledge of the statistical distributions for the different ensembles is sufficient to characterize the macroscopic properties of a physical system in thermal equilibrium. We now make the link explicit between the statistical description and the expressions of macroscopic thermodynamics, for which a number of basic principles can be formulated.

The Laws of Thermodynamics. The physical insight of macroscopic thermodynamics is usually summarized in the *laws of thermodynamics*. They are

- Law **0**: If two systems A and B are in *thermodynamic equilibrium*, and B is in equilibrium with C, then A and C are also in equilibrium with each other. The equilibria can be characterized by their mechanical, thermal or chemical properties.

- Law **1**: Energy E is conserved:

$$dE = dQ + dW \qquad (1.12)$$

where Q is the *heat* flowing into the system, and W the *work* done by the system. If a process is *adiabatic*, dQ is zero.

- Law **2**: In a thermally isolated macroscopic system, entropy never decreases.

---

[5]The chemical potential often confuses: it is a measure for the availability of particles, and plays thus an analogous role as temperature does in providing an energy source.

[6]There are exceptions for the equivalence of ensembles, a point we do not pursue here. And, obviously, for systems in which N is *very much smaller* than Avogadro's constant, the deviations from the thermodynamic limit can be important.

- Law **3**: As the temperature of a system tends to absolute zero, entropy reaches a constant.

We see that energy and entropy are the key notions of equilibrium thermodynamics. Let us understand them more deeply. To achieve this we first generalize the descriptors we have introduced before and introduce the important notion of an *extensive* variable.

An extensive variable is a variable which scales linearly with system size. Volume is a (trivial) example for such a system property: if one starts with two systems of volume V_1 and V_2, the volume of the systems is additive: $V = V_1 + V_2$, and hence scales linearly. The particle number N is another example. Temperature T, however, is not extensive: putting two systems of equal temperature T together does not result in a system with temperature $2T$.

We call the state of a system an *equilibrium state* if it can be characterized by its (internal) energy E, and a set of extensive parameters $X_1, ... X_m$. For such a state there exists a function of the extensive parameters, the entropy S (which we obtained before from the Jaynes principle). We call the relation

$$S = S(E, X_0, ..., X_m) \tag{1.13}$$

a fundamental equation for the system. S itself is also extensive; furthermore, it is an increasing function[7] of E.

Consider now the first statement, the notion of the extensivity of S. We can express it mathematically by writing S as a *first-order homogeneous* function of its extensive parameters

$$S(\lambda E, \lambda X_0, ... \lambda X_m) = \lambda S(E, X_0, ..., X_m), \tag{1.14}$$

where λ is a scalar.

The second statement about S, the monotonicity property, corresponds to the condition[8]

$$\left(\frac{\partial S}{\partial E} \right)_{X_i} \geq 0. \tag{1.15}$$

---

[7] E is also extensive, but this is actually a subtle point if systems become strongly interacting. We are not concerned with sophisticated problems of this sort here.
[8] The equality is restricted to a set of measure zero.

The monotonicity property of S with respect to E allows its inversion, leading to

$$E = E(S, X_0, ..., X_m) \qquad (1.16)$$

and E, like S, is a first-order homogeneous function

$$E(\lambda S, \lambda X_0, ...\lambda X_m) = \lambda E(S, X_0, ..., X_m). \qquad (1.17)$$

Some properties of scalar fields. Before introducing the thermodynamic potentials, we need some mathematical concepts for scalar fields. They are stated here just as facts.

- If $\phi = \phi(x_0, ..., x_m)$ is a scalar field of $m+1$ variables, its *total differential* is given by

$$d\phi = \sum_{i=0}^{m} \frac{\partial \phi}{\partial x_i} dx_i. \qquad (1.18)$$

- If the $x_i = x_i(u, v)$ with u, v scalar, we can rewrite this expression

$$d\phi = \sum_{i=0}^{m} \frac{\partial \phi}{\partial x_i} \frac{\partial x_i}{\partial u} du + \sum_{i=0}^{m} \frac{\partial \phi}{\partial x_i} \frac{\partial x_i}{\partial v} dv. \qquad (1.19)$$

- *Contour surfaces*, for $\phi = constant$, define an implicit functional relationship between the x_i since on a contour surface

$$\sum_{i=0}^{m} \frac{\partial \phi}{\partial x_i} dx_i = 0. \qquad (1.20)$$

- If all x_i except x_0 and x_1 (e.g.) are held fixed, then

$$\left(\frac{\partial x_1}{\partial x_0} \right)_{\phi,...} = \left(\frac{\partial \phi}{\partial x_0} \right)_{x_1,...} \cdot \left(\frac{\partial \phi}{\partial x_1} \right)_{x_0,...}^{-1}, \qquad (1.21)$$

and

$$\left(\frac{\partial x_0}{\partial x_1} \right)_{\phi,...} = \left(\frac{\partial x_1}{\partial x_0} \right)_{\phi,...}^{-1} \qquad (1.22)$$

for $\phi = const.$

- For three variables, one has the cyclic rule

$$\left(\frac{\partial x_0}{\partial x_1} \right)_{x_2} \cdot \left(\frac{\partial x_1}{\partial x_2} \right)_{x_0} \cdot \left(\frac{\partial x_2}{\partial x_0} \right)_{x_1} = -1, \qquad (1.23)$$

which generalizes to more variables in an obvious way.

Exercise. Perform the generalization of the cyclic rule, Eq. (1.23), to more variables.

We can now proceed to look at the total differentials of E and S. We assume that E is a function of entropy S, volume V, and particle number N; the extension to further extensive variables is straightforward. We have

$$dE = \left(\frac{\partial E}{\partial S}\right)_{V,N} dS + \left(\frac{\partial E}{\partial V}\right)_{S,N} dV + \left(\frac{\partial E}{\partial N}\right)_{S,V} dN \qquad (1.24)$$

where the subscripts indicate the variables to be held fixed. We define the following intensive parameters:

$$T = \left(\frac{\partial E}{\partial S}\right)_{V,N} \qquad (1.25)$$

is *temperature*;

$$P = -\left(\frac{\partial E}{\partial V}\right)_{S,N} \qquad (1.26)$$

is *pressure*;

$$\mu = \left(\frac{\partial E}{\partial N}\right)_{S,V} \qquad (1.27)$$

is *chemical potential*; hence

$$dE = TdS - PdV + \mu dN . \qquad (1.28)$$

The intensive parameters are all functions of S, V, N, and the functional relationships in eqs.(1.25) - (1.27) are called *equations of state*.

Exercise. Verify explicitly that temperature T, pressure P and chemical potential are intensive variables, i.e., that they are *homogeneous functions of zeroth order*.

Task. Deduce from the fact that E is a first-order homogeneous function of its variables one can obtain the *Euler equation.*[9]

$$E = TS - PV + \mu N . \qquad (1.29)$$

---

[9]The expressions for E show that TS has the dimension of an energy. Hence the information entropy we determined in the first section has to be multiplied by Boltzmann's constant, k_B.

Hint: differentiate Eq. (1.17) with respect to λ and put $\lambda = 1$.

Given that we have introduced E as a function of entropy S, volume V and particle number N, one may wonder whether it is not possible - and even desirable - to define thermodynamic functions of other variables than the ones chosen. In particular, one may also want to study dependences on intensive rather than extensive variables. This is indeed possible, and the resulting functions are not independent from each other. In fact, one can pass from one to the other via a mathematical transformation we introduce first in a formal manner.

Legendre transform. Let $Y(X_0, ..., X_m)$ be a scalar field of the extensive variables X_j, and the $P_j = (\partial Y/\partial X_j)_{X_{i \neq j}}$ are the corresponding intensive variables. Then

$$\Lambda = Y[P_0, ..., P_i]_{i \leq m} \equiv Y - \sum_{j=0}^{i} X_j P_j \tag{1.30}$$

is the *Legendre transform* of Y with respect to $X_{j \leq i}$. The total differential of Λ reads

$$d\Lambda = -\sum_{j=0}^{i} X_j dP_j + \sum_{j=i+1}^{m} P_j dX_j \tag{1.31}$$

so that $\Lambda = \Lambda(P_0, ..., P_i, X_{i+1}, ..., X_m)$ is a function of the $i+1$ intensive, and $m-i$ extensive variables.

After this formal definition we want to apply this concept. This is best done by simple cases, ignoring the physical context for the moment.

Legendre transform in one dimension. We first discuss the Legendre transform in one dimension for a function $f(x)$ and its derivative

$$y = f'(x) = \frac{df}{dx} \equiv g(x). \tag{1.32}$$

We can understand $y = g(x)$ as a variable transformation from x to y. How can we express f in terms of y?

For the function $f(x)$ we can write the equation within each point x along the curve

$$f(x) = xf'(x) + b(x). \tag{1.33}$$

Since $x = g^{-1}(y)$ where g^{-1} is the inverse of g, we have

$$b(g^{-1}(y)) = f(g^{-1}(y)) - yg^{-1}(y) \equiv \Lambda(y). \tag{1.34}$$

This is often written in short form as

$$\Lambda(y) = f(x) - xy, \tag{1.35}$$

where x is a function of y.

We move on to some exercises.

Exercise. Compute the Legendre transform of $f(x) = e^x$.

Exercise. Give the Legendre transform of the harmonic form with the vector $\mathbf{x}$,

$$u(\mathbf{x}) = \frac{1}{2}(\mathbf{x}^T \cdot \mathbf{A} \cdot \mathbf{x}), \tag{1.36}$$

where $\mathbf{A}$ is an invertible $(n \times n)$-matrix.

We now return to the physical context. For the energy E, the four most common thermodynamic potentials that result from the application of the Legendre transform are the

- *Helmholtz free energy* (T given, canonical ensemble)

$$F(T, V, N) = E[T] = E - TS = -PV + \mu N \tag{1.37}$$

$$dF = -SdT - PdV + \mu dN \tag{1.38}$$

- *Enthalpy* (P given)

$$H(S, P, N) = E[P] = E + PV = TS + \mu N \tag{1.39}$$

$$dH = TdS + VdP + \mu dN \tag{1.40}$$

- *Gibbs-free energy* (T, P given)

$$G(T, P, N) = E[T, P] = E + PV - TS = \mu N \tag{1.41}$$

$$dG = -SdT + VdP + \mu dN \tag{1.42}$$

- *Grand canonical potential* (T, μ given)

$$\Phi(T, V, \mu) = E[T, \mu] = E - TS - \mu N = -PV \tag{1.43}$$

$$d\Phi = -SdT - PdV - Nd\mu \tag{1.44}$$

The thermodynamic potentials all have to be minimized at equilibrium for fixed values of their variables.

This ends our brief look into thermodynamics. The message that we want to retain is that

- within equilibrium physics there is a well-established body of thermodynamic functions with which the macroscopic properties of a system can
  be described;

- these functions have a rigorous link with each other via the Legendre
  transform, and with statistical mechanics, since we know how to compute
  the thermodynamic potentials within this theory.

We now return to statistical mechanics, and begin to discuss methods which
allow to compute the partition function and the quantities derivable from it.

1.3 COMPUTING Z

This section introduces methods to compute Z and the thermodynamic quantities that can be derived from it. We begin with some technicalities that will
be useful later. An obvious first step in the computation of Z is to know how
to compute integrals involving the *Gaussian distribution*. The Gaussian distribution is a generic characteristic of equilibrium states: in thermal equilibrium,
a system will be in a state minimizing the corresponding thermodynamic potential, and the fluctuations around this stable state will be Gaussian.

Gaussian distribution. The Gaussian probability distribution in the case
of one variable $-\infty < x < \infty$ is given by

$$P(x) = Ce^{-\frac{1}{2}Ax^2 - Bx} \qquad (1.45)$$

where the normalization constant is

$$C = \left(\frac{A}{2\pi}\right)^{1/2} e^{-\frac{B^2}{2A}}. \qquad (1.46)$$

The parameter $A > 0$ controls the width of $P(x)$ and, together with B, the
peak position. Introducing the average μ and variance σ^2 we find

$$\mu_1 = -\frac{B}{A}, \quad \sigma^2 = \frac{1}{A} \qquad (1.47)$$

and can thus write the normalized form of the Gaussian or *standard normal
distribution*

$$P(x) = \frac{1}{\sqrt{2\pi\sigma^2}} e^{-\frac{(x-\mu_1)^2}{2\sigma^2}}. \qquad (1.48)$$

The *multivariate* version of the distribution, for $i = 1, ..., m$ random variables $x = \{x_i\}$ is

$$P(x) = C \exp\left[-\frac{1}{2}\sum_{i,j=1}^{m} A_{ij}x_i x_j - \sum_{i=1}^{m} B_i x_i\right] \tag{1.49}$$

where $\mathbf{A} = A_{ij}$ is a positive definite symmetric matrix. The normalization constant reads in this case as

$$C = (2\pi)^{-m/2}(Det\,\mathbf{A})^{-1/2}\exp\left[-\frac{1}{2}\mathbf{B}\cdot\mathbf{A}^{-1}\cdot\mathbf{B}\right] \tag{1.50}$$

with the matrix inverse $\mathbf{A}^{-1}$.

Based on these results we can now easily write down the *mean* of each of the x_i

$$\langle x_i \rangle = -\sum_j (A^{-1})_{ij} B_j \tag{1.51}$$

and the *covariance*

$$\langle (x_i - \langle x_i \rangle)(x_j - \langle x_j \rangle) \rangle = \langle x_i x_j \rangle - \langle x_i \rangle\langle x_j \rangle = (A^{-1})_{ij}. \tag{1.52}$$

The inverse of A is the *covariance matrix*. Its diagonal elements are the variances, the off-diagonals are the covariances.

Characteristic function, moment generating function and cumulants. The *characteristic function* of a *stochastic variable* X is defined by

$$G(k) \equiv \langle e^{ik} \rangle = \int dx P(x) e^{ikx}, \tag{1.53}$$

where the symbol $\langle ... \rangle$ was used to abbreviate the average in distribution. Obviously, $G(k)$ is a Fourier transform of P. It exists for real k and obeys

$$G(0) = 1, \quad |G(k)| \leq 1. \tag{1.54}$$

The coefficients μ_m of its Taylor expansion are the *moments*

$$G(k) = \sum_{m=0}^{\infty} \frac{(ik)^m}{m!}\mu_m \tag{1.55}$$

where

$$\mu_n = \langle x^n \rangle = \int dx P(x) x^n. \tag{1.56}$$

The coefficients κ_m of the series of its logarithm,

$$\ln G(k) = \sum_{k=1}^{\infty} \frac{(ik)^m}{m!} \kappa_m \tag{1.57}$$

are the *cumulants*. They are combinations of the moments, the lowest of which are

$$\kappa_1 = \mu_1, \quad \kappa_2 = \mu_2 - \mu_1^2 = \sigma^2. \tag{1.58}$$

For the Gaussian distribution, all higher cumulants vanish. Higher moments can thus serve to characterize more complex distributions - this is useful since not in all cases full distributions are obtainable.

Exercise. Compute the cumulants for the *Poisson distribution*

$$p_n = \frac{a_n}{n!} e^{-n} \tag{1.59}$$

over the integers and zero, $n = 0, 1, 2, \dots$.

We have reached a point where we want to see the machinery of statistical mechanics in action. We will now apply it to a standard model of statistical mechanics, the *Ising model*, certainly the most famous model ever formulated in statistical mechanics. Originally conceived to describe a uniaxial ferromagnet and its phase transition from the paramagnetic (non-magnetic) to the ferromagnetic phase, it has found innumerable applications all over the fields of physics, chemistry and biology.

1.4 THE ISING MODEL

The Ising model on a one-dimensional lattice is defined by the energy[10]

$$H = -\frac{K}{2} \sum_n s_n s_{n+1} - h \sum_n s_n \tag{1.60}$$

where the $s_n = \pm 1$ are called the 'spins' - local magnetic moments - of a ferromagnet.[11] Within the model, the parameter K describes a coupling between neighbouring local spins; for $K > 0$ they will prefer to be in the same

---

[10]In statistical mechanics jargon, the *Hamiltonian*.

[11]A ferromagnet is usually a metal which, below a certain temperature, has a permanent magnetic moment; the origin of this permanent moment lies in the microscopic magnetic moments produced by the electron shells inside the metal. In the model description we employ, we summarize these atomistic effects by an elementary spin. This is sufficient if we

state since the energy favors that state. The parameter h in the second term defines an external coupling affecting each of the spins; for the ferromagnetic system for which the Ising model was originally conceived, this contribution represents an applied magnetic field which, depending on its sign, can favour either the 'up' or the 'down' state with all spins following the field direction.

Since we will use the Ising model in a biological context, it is clear that the original interpretation of the parameters is largely immaterial for us: for us, s_n is just a two-valued variable, and can be re-interpreted at will. From a biological point of view we can, e.g., consider the Ising model in one spatial dimension as a chain of objects which can be in either of two states. A simple biological example to which this model then applies can be a DNA molecule. Within the Ising model we can define microstates that distinguish each other by the different numbers of spins up, interpreted as 'base pairs bound' and spins down, interpreted as 'base pairs unbound'. Hence we can, e.g., address the question of the binding of two DNA strands, and the Ising model becomes a first - but ultimately too crude - model for the physics of a DNA double strand.

This re-interpretability of the model is a big advantage of statistical mechanics: indeed, very often physically quite different situations turn out to fall into the class of just one type of model. This explains the success of the Ising model in so many applications - having a system with just two alternatives is the simplest one can have.

Plugging the Hamiltonian of the Ising model into the partition function in the canonical ensemble, we obtain

$$Z = \sum_{\{s_n\}} e^{-H} = \sum_{\{s_n\}} e^{J \sum_n s_n s_{n+1} + \beta h \sum_n s_n} \tag{1.61}$$

with $J \equiv \beta K/2$. In the following we leave out the field h; its inclusion in the calculation is left as a *Task*.

Now we have to compute Z, and we will learn a first method to do it.

The transfer-matrix solution of the 1d-Ising model. The partition function can be rewritten as

$$Z = \sum_{\{s_n\}} e^{J s_0 s_1} \cdot e^{J s_1 s_2} \cdot \ldots \cdot e^{J s_{N-1} s_0} \tag{1.62}$$

---

are only interested in the ordering phenomenon between a paramagnetic (non-magnetic) and a ferromagnetic phase. A true microscopic theory of ferromagnetism has to be a quantum theory.

where a *periodic boundary condition* is assumed, i.e., the linear chain is joined at its ends.

Since each spin has two orientations, there are four configurations for two neighbouring spins, two of which are degenerate in energy. This suggests to introduce a matrix representation with a *transfer matrix*

$$T_J = \begin{pmatrix} e^J & e^{-J} \\ e^{-J} & e^J \end{pmatrix} \tag{1.63}$$

which is a symmetric matrix, to be diagonalized by an orthogonal matrix O. Since this matrix applies to every neighbouring pair of spins we can express the partition function as

$$Z = Tr[T_J^N] = Tr[(OT_JO^{-1})^N] \tag{1.64}$$

where the symbol Tr stands for 'Trace', the sum of the diagonal entries of the matrix.

Exercise. Verify expression (1.64).

The orthogonal matrix O to diagonalize T_J is given by

$$O = \begin{pmatrix} 0 & 1 \\ -1 & 0 \end{pmatrix} \tag{1.65}$$

and we find for the diagonalized matrix the result

$$OT_JO^{-1} = 2 \begin{pmatrix} \cosh J & 0 \\ 0 & \sinh J \end{pmatrix} \tag{1.66}$$

so that

$$Z = 2^N[\cosh^N J + \sinh^N J]. \tag{1.67}$$

Since in statistical mechanics we are interested in the thermodynamic limit which in this model amounts to consider $N \to \infty$, one finally finds the expression of the partition function

$$Z = 2^N \cosh^N J. \tag{1.68}$$

Exercise. Show (1.68).

Eq. (1.68) is the first partition function we have calculated explicitly, and we can now calculate physical quantities from it. Let us begin with the internal energy E; it is found to be

$$E = Z^{-1} \sum_{\{s_n\}} He^{-\beta H} = Z^{-1} \frac{\partial Z}{\partial \beta} = -NK \tanh(\beta K/2). \qquad (1.69)$$

A second, more complex example is the *two-point correlation function* $\langle s_i s_j \rangle$ for $|i - j| = r$. It is given by

$$\langle s_i s_j \rangle = Z^{-1} \sum_{\{s_n\}} s_i s_j e^{-\beta H}. \qquad (1.70)$$

In order to apply the transfer matrix method we first have to find a suitable way to represent it in terms of T_J. One can convince oneself that the expression to write down is

$$\sigma_3 T_J^r \sigma_3 T_J^{N-r} \qquad (1.71)$$

where σ_3 is the matrix

$$\sigma_3 = \begin{pmatrix} 1 & 0 \\ 0 & -1 \end{pmatrix}. \qquad (1.72)$$

This expression states that between two well-defined spin states along the chain the transfer matrix has to propagate r times between the spins, and then again $N - r$ times along the other side of the closed chain - remember we maintain the periodic boundary condition.

With this we can now write

$$\begin{aligned}
\langle s_i s_j \rangle &= Z^{-1} Tr[\sigma_3 T_J^r \sigma_3 T_J^{N-r}] \\[2mm]
&= Z^{-1} Tr[(O\sigma_3 O^{-1}) \cdot (OT_J O^{-1})^r \cdot (O\sigma_3 O^{-1}) \cdot (OT_J O^{-1})^{N-r}] \\[2mm]
&= 2^N Z^{-1} (\cosh^r J \sinh^{N-r} J + \sinh^r J \cosh^{N-r} J) \\[2mm]
&= \tanh^{N-r} J + \tanh^r J \qquad (1.73)
\end{aligned}$$

where

$$O\sigma_3 O^{-1} = \begin{pmatrix} 0 & 1 \\ 1 & 0 \end{pmatrix}. \qquad (1.74)$$

With $N \gg r$ and in the thermodynamic limit we obtain the expression

$$\langle s_i s_j \rangle = \tanh^r J = e^{-|i-j|/\xi} \tag{1.75}$$

where we defined

$$\xi \equiv |\ln(\tanh J)|^{-1} \tag{1.76}$$

as the *correlation length*. For the 1d-Ising model ξ diverges in the limit $T \to 0$, when all spins align either up or down (i.e., the system adopts a homogeneous, oriented state). It will turn out that there is no *phase transition* in the one-dimensional Ising model at a finite temperature ($T > 0$) between a state of finite magnetization - the oriented state in which all spins point collectively either up or down - and a state without a net magnetization, in which the spins point randomly either up or down with a vanishing average.

For this simple example, we are essentially done and the reader can try to extend the model to include the field h, or calculate other thermodynamic quantities to get a feeling for the systems' properties.

But we do not yet want to stop here and take at little deeper look into this system. We have computed thermodynamic quantities in the thermodynamic limit $N \to \infty$. Let us suppose we do not do that - in fact, in some applications of statistical mechanics in biology, this limit may be difficult to reach, and effects of finite system size may be important.

Let us therefore study how the properties of the one-dimensional Ising model are affected upon a change of N. We start from the expression

$$Z[N, J] = Tr[T_J^N] = Tr[(OT_JO^{-1})^N] \tag{1.77}$$

where $OT_JO^{-1} = e^J \cdot \mathbf{1} + e^{-J}\sigma_3$; the N-dependence is now explicit. Suppose we, in a first step, double the lattice size from N to $2N$. With

$$(OT_JO^{-1})^2 = 2(\cosh 2J \cdot \mathbf{1} + \sigma_3) = COT_{\bar{J}}O^{-1} \tag{1.78}$$

where

$$C = 2\sqrt{\cosh 2J}, \quad \bar{J} = \frac{1}{2}\ln(\cosh 2J) \tag{1.79}$$

we arrive at the relation

$$Z[2N, J] = C^N Z[N, \bar{J}]. \tag{1.80}$$

The equations (1.79) and (1.80) can be read as *generators of a flow* in the coupling constant J. If we were to iterate the mapping $J \to \bar{J}$ starting from a value $J = 1$, the iterated map will converge to zero (*Exercise*). What can we learn from this observation?

By these very simple considerations we have in fact found the two *fixed points* of a *renormalization group transformation* of the 1d-Ising model - without noticing that such a thing exists. The first fixed point is $J = 0$, which corresponds to $T = \infty$. Physically, the system is then in the fully disordered phase, in which the spins point randomly either up or down. The second fixed point is $J = \infty$ corresponding to $T = 0$, and this is the fully ordered spin state which only exists at zero temperature. Indeed, there is no phase with a finite magnetization $M \equiv \langle s_i \rangle \neq 0$ for $0 < T \leq \infty$, as we mentioned before.

Although we have of course not made a systematic approach to renormalization, there is already something to learn. Suppose we consider the system at a given temperature, and at a given size N. The spins inside the system will have a correlation length ξ over which their orientation (up or down) is correlated. If we now double the system size, what does the correlation do? Will it grow stronger or not? It is characteristic of a system *right at a phase transition* that the correlation length is infinite. The system behaves in a collective (or cooperative) fashion. If the temperature I choose is just that critical temperature, all the steps of increasing system size will not affect the correlation length, and it will stay infinite. For the 1d-Ising model, this is just the case at $T = 0$. By contrast, if I deviate ever so slightly from that critical temperature, the step of increasing system size will tend to reduce the correlation length, and if this procedure is repeated over and over again[12] the system will move away to a system of uncorrelated spins, hence a disordered phase. This whole approach, in its systematic version, allows to compute the properties of all equilibrium phase transitions in a unified way.

After this digression on the idea behind the renormalization group, we turn to another approach to compute the partition function of the 1d-Ising model: the computation by recursion. This will turn out to be a very important technique in the third chapter of this book, since recursion methods are at the heart of dynamic programming.

---

[12]Hence the notion of a renormalization *group*; technically, it is a semi-group.

Solution of the 1d-Ising model by recursion. The analytic solution of the 1d-Ising model is so easy to obtain since the coupling between any two spins is the same. This is not necessarily the case, and we will indeed see later that for the most relevant biological applications this simplifying assumption cannot be made.

So let's now make things a little more complicated. For the 1d-Ising model with a *neighbour-dependent coupling* $J(n)$, we write[13]

$$\beta H = -\sum_{n=1}^{N-1} J(n)s(n)s(n+1) - \sum_{n=1}^{N} H(n)s(n) \tag{1.81}$$

where we now have also included a neighbour-dependent field $H(n) = \beta h(n)$. This variant of the Ising-model has been applied to problems in statistical genetics, as suggested by J. MAJEWSKI et al., 2001, in a study of *epistasis* (i.e., gene-gene interaction). In this interpretation, the spins on the lattice are identified with different genes placed along the DNA molecule rather than the individual base pairs.

As before, the total number of states in the model is 2^N, hence exponential in N, and the computation to be performed is seemingly exponential in N since we now cannot use the transfer matrix trick anymore due to the neighbour-dependence of couplings. Using a *recursion technique*, however, we can achieve a computation in linear 'time' N.

For the construction of the recursion relation we write

$$Z_N = Z_{N+} + Z_{N-} \tag{1.82}$$

where Z_{N+} refers to the partition function of the chain (which is left open in this case - in contrast to our computation before) in which the last spin at site N points up; the interpretation of Z_{N-} is evident. One finds (*Exercise*)

$$Z_{N+} = e^{H(N)}[Z_{(N-1)+}e^{J(N-1)} + Z_{(N-1)-}e^{-J(N-1)}] \tag{1.83}$$

and

$$Z_{N-} = e^{-H(N)}[Z_{(N-1)+}e^{-J(N-1)} + Z_{(N-1)-}e^{J(N-1)}]. \tag{1.84}$$

The partition function Z_N can now be computed recursively from these expressions.

This concludes the discussion of the Ising model in one spatial dimension.

---

[13]With a slight change of notation which should be obvious.

Although we saw some computational approaches at work, we can be a little bit dissatisfied since there is no phase transition between an ordered and a disordered state in this model at a finite temperature. So we are asked to generalize, and one way to do this is to pass on to higher dimensions.[14] Since calculations in higher dimensions are more complicated (and in fact, they will become far too complicated for the ambitions of this text) we will try to simplify in another way. This leads us to what is called the *mean-field solution* of the Ising model. In this approach, spatial dimension will first play no role.

Mean-field solution. For the discussion of the *mean-field approximation* we place the spins on the edges of a hypercube in d space dimensions, and write the energy as

$$H = -\frac{1}{2} \sum_{\mathbf{x}, \mathbf{y}} J(\mathbf{x}, \mathbf{y}) s(\mathbf{x}) s(\mathbf{y}) \qquad (1.85)$$

where $\mathbf{x} = a\, x_i\, \mathbf{e}_i$ with the lattice constant a and $\mathbf{e}_i$, $i = 1, ..., d$ are the unit vectors on the lattice. $J(\mathbf{x}, \mathbf{y})$ is taken as a symmetric matrix.

Again we need a technical concept first.

Hubbard-Stratonovich transformation. Consider the Gaussian integral relation

$$\exp\left(\frac{\beta}{2} J s^2\right) = \sqrt{\frac{\beta}{2\pi J}} \int_{-\infty}^{\infty} d\phi \exp\left(-\frac{\beta}{2} J^{-1} \phi^2 + \beta \phi s\right). \qquad (1.86)$$

It can be understood as a *linearization* of the argument of the exponential (i.e., the variable s). The price to pay for this operation is the introduction of an auxiliary integration variable.

We generalize this step to the weight

$$\exp\left(\frac{\beta}{2} \sum_{\mathbf{x}, \mathbf{y}} J(\mathbf{x}, \mathbf{y}) s(\mathbf{x}) s(\mathbf{y})\right) = \qquad (1.87)$$

$$= \prod_{\mathbf{x}} \int_{-\infty}^{\infty} d\phi(\mathbf{x}) \exp\left[-\frac{\beta}{2} \sum_{\mathbf{x}, \mathbf{y}} J^{-1}(\mathbf{x}, \mathbf{y}) \phi(\mathbf{x}) \phi(\mathbf{y}) + \beta \sum_{\mathbf{x}} \phi(\mathbf{x}) s(\mathbf{x})\right]$$

where we now have introduced a local *field* $\phi(\mathbf{x})$ at each lattice site. Note that

---

[14]With this generalization we lose for the moment the biological interpretation of the Ising model for the binding of a DNA double strand. But this does not matter: here we are mainly interested in the problem of the calculation of partition functions; the more specific use of such models within a given biological context follows in the subsequent parts of the book.

the expression is meaningful as long as the matrix is semi-definite and positive, which is not the case for the Ising model since $J(\mathbf{x}, \mathbf{x}) = 0$; the formula thus has to be considered a formal one.[15]

The partition sum over the spins s can be performed exactly with the *Hubbard-Stratonovich transformation*, with the result

$$Z = \int D\phi \, \exp\left[-\frac{\beta}{2}\sum_{\mathbf{x},\mathbf{y}} J^{-1}\phi(\mathbf{x})\phi(\mathbf{y}) + \sum_{\mathbf{x}} \ln(\cosh[\beta\phi(\mathbf{x})])\right] \qquad (1.88)$$

where the integral measure is $\int D\phi \equiv \int \prod_{\mathbf{x}} d\phi(\mathbf{x})$. The $\ln(\cosh(..))$-term in the argument of the exponential function arises from the summation over the spins $s(\mathbf{x})$ which is now easy to do - that was just the idea behind the linearization in the first place.

We now have the expressions in place to perform the mean-field approximation. It corresponds to taking the *saddle-point* of the integrand, which is determined by the stationary value of the argument of the exponential. This idea is easily explained for a function of one variable; it is also known as the *method of steepest descent*.[16] Suppose the integral we want to compute is

$$I = \lim_{N\to\infty} \int_{-\infty}^{\infty} dx \, e^{-Nf(x)} \qquad (1.89)$$

where we have introduced an explicit parameter N; frequently such a parameter can be defined, if often only on formal grounds. Suppose the function f has a global minimum at a value $x = x_0$, well-separated from possibly other minima. Upon Taylor expansion, f fulfills

$$f(x) \approx f(x_0) + \frac{1}{2}f''(x_0)(x - x_0)^2 + \mathcal{O}(x^3), \qquad (1.90)$$

since we expand around an extremum for which $f'(x_0) = 0$.

In the limit of $N \to \infty$, the global minimum will dominate the integrand and the integration range is largely determined by the region around x_0. Hence

$$I \approx \lim_{N\to\infty} e^{-Nf(x_0)} \int_{-\infty}^{\infty} dx \, e^{\frac{N}{2}f''(x_0)(x-x_0)^2} \qquad (1.91)$$

$$\approx \lim_{N\to\infty} e^{-Nf(x_0)} \left(\frac{2\pi}{Nf''(x_0)}\right)^{1/2}.$$

---

[15] Further, we have ignored the integration amplitude.
[16] For oscillating integrands, see the *stationary phase approximation*.

Task. Compute the next-order correction to the saddle-point value of the integral.

Coming back to our case, the result is given by the equation

$$-\sum_{\mathbf{y}} J^{-1}(\mathbf{x}, \mathbf{y})\phi_0(\mathbf{y}) + \tanh[\beta\phi_0(\mathbf{x})] = 0. \qquad (1.92)$$

with the mean-field $\langle\phi(\mathbf{x})\rangle = \phi_0(\mathbf{x})$.

A note on the mean-field approximation. At this point it is useful to make a comment on the notion of the mean-field approximation since the approach we took seems rather technical. In fact, there are several ways of introducing a mean-field. A simple alternative is, e.g., to define and introduce

$$\widehat{H}(\mathbf{x}) \equiv \sum_{\mathbf{y}} J(\mathbf{x}, \mathbf{y})s(\mathbf{y}) \qquad (1.93)$$

as an effective field acting on the spins $s(\mathbf{y})$. The resulting partition function then becomes one-dimensional and can be solved following the steps we took before. This procedure essentially means that within mean-field theory, one neglects correlations and factorize $\langle s(\mathbf{x})s(\mathbf{y})\rangle = \langle s(\mathbf{x})\rangle\langle s(\mathbf{y})\rangle$. While this procedure is technically much easier to perform than the computation we did, we will see in the following that our more systematic and general approach will also allow us, in a fairly straightforward sequence of steps, to learn something about the regime of validity of the mean-field approximation.

We now return to Eq. (1.92). If we assume that $J(\mathbf{x}, \mathbf{y}) \equiv J$, the system becomes translationally invariant and we can define the uniform *order parameter M* via

$$\phi_0 = 2dJM \qquad (1.94)$$

so that the saddle-point equation is rewritten as

$$M = \tanh(2d\beta JM). \qquad (1.95)$$

The solutions of this equation can be obtained graphically, see Figure 1.1. If $2d\beta JM > 1$, there is a pair of solutions which merge at $M = 0$ at the critical temperature

$$\beta_c^{-1} = 2dJ. \qquad (1.96)$$

Within the mean-field approximation, the Ising model thus displays a phase transition at a finite temperature. This is at variance with our exact result for $d = 1$. So we have to clarify the limit of validity of the approximation we made.

In fact, as we said before, the mean-field approximation assumes a factorization of correlation functions, hence a neglect of fluctuations.[17] These can be accounted for by an expansion around the saddle-point value. We may thus speculate that the approximation is valid whenever this expansion is meaningful. In order to approach this question, we therefore have to pass on to a theory which is capable to capture the properties of the Ising model in the vicinity of the critical temperature.

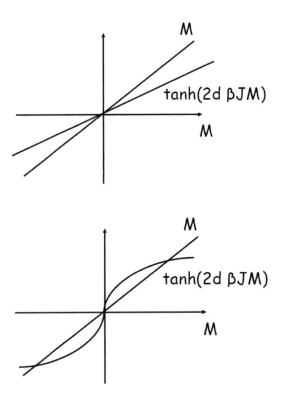

Figure 1.1 Phase transition in the Ising model in the mean-field approximation: graphical solution of the mean-field equation. A change in temperature allows to pass from a unique solution for $T > T_c$ (left) to two solutions at $T < T_c$ (right).

---

[17] Why do fluctuations destroy the factorization of correlation functions? *Exercise.*

Ginzburg-Landau theory. In the vicinity of the transition, we can expand the terms in Eq. (1.88) in a Taylor series (Why? *Exercise!*)

$$\sum_{\mathbf{x},\mathbf{y}} J^{-1}\phi(\mathbf{x})\phi(\mathbf{y}) = J^{-1}\sum_{\mathbf{x}} \phi(\mathbf{x})\left(\frac{1}{2d} - \frac{1}{4d^2}a^2\nabla^2 + ...\right)\phi(\mathbf{x}) \qquad (1.97)$$

and

$$\ln[\cosh(\beta\phi(\mathbf{x}))] = \frac{\beta^2}{2}\phi^2(\mathbf{x}) - \frac{\beta^4}{12}\phi^4(\mathbf{x}) + ... \qquad (1.98)$$

which, after passing to the continuum limit for $a \to 0$ yields the *Ginzburg-Landau* form of the Hamiltonian

$$\beta H = \int d^d\mathbf{x}\left[\frac{1}{2}(\nabla\phi)^2 + \frac{m^2}{2}\phi^2 + \frac{\lambda}{4!}\phi^4\right]. \qquad (1.99)$$

This theory can easily be treated. We consider the case of a homogeneous order parameter, $\phi = const$. Then

$$\beta V_0(\phi) = \frac{m^2}{2}\phi^2 + \frac{\lambda}{24}\phi^4 \qquad (1.100)$$

is the *mean-field* or *effective potential* of the Ginzburg-Landau theory. Depending on the sign of the quadratic term - with a fourth-order term which has to be strictly positive for thermodynamic stability - the potential displays two shapes which are shown in Figure 1.2. For $m^2 > 0$, the potential has a single minimum at $\phi = 0$; this can be identified with a disordered state. For $m^2 < 0$, two minima appear at values $\phi = \pm\phi_0$. They correspond to a pair of ordered states related to each other by a mirror-symmetry: one has a positive, the other a negative value of the same magnitude. For the ferromagnet these states correspond to states with positive and negative magnetization and are identical to the solutions obtained from the graphical solution of the saddle-point equation in Figure 1.1.

So far we have not specified the value of m^2; in any case, in the vicinity of the transition at $m = 0$ we know that

$$m^2 \sim \beta_c - \beta \sim |T - T_c|. \qquad (1.101)$$

On the other hand, on dimensional grounds, m must be an *inverse length*, and we define

$$m \equiv \xi^{-1} \sim |T - T_c|^{1/2}, \qquad (1.102)$$

i.e.,

$$\xi \sim |T - T_c|^{-1/2}. \qquad (1.103)$$

We had encountered ξ before: it is the (spin-spin) correlation length. In Eq. (1.103) we have obtained the first *critical exponent* which describes the power-law behaviour of physical quantities in the vicinity of the critical point: the exponent $\nu = 1/2$ is thus the critical exponent of the correlation length.

The information that the approach to a critical point is by a power-law is essential. Power laws are a general characteristic of systems without an intrinsic length scale (an example is a decay-length of correlations), or, in other words, in *scale-invariant* systems. This property of critical systems explains why power laws are so dearly loved by statistical physicists, and why they try to find them in more complex systems as well - a point we will return to in the book.

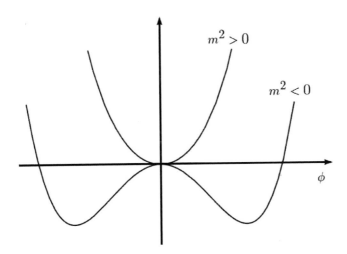

Figure 1.2 Mean-field potential of the Ginzburg-Landau theory, for $T < T_c$ ($m^2 < 0$) and $T > T_c$ ($m^2 > 0$).

The Ginzburg-Landau model is, from a computational point of view, a very nice topic. We therefore suggest a number of exercises and tasks for the readers to try.

Task. Perform the steps explicitly which are needed to pass from the discrete to the continuum version of the Ginzburg-Landau theory, leading to Eq. (1.99). Verify that the mass parameter and the coupling constant are given by

$$m^2 = \frac{\beta_c^{-2}}{Ja^2}(\beta_c - \beta), \quad \lambda = \frac{4\beta^2}{J^2\beta_c^4}a^{d-4}. \tag{1.104}$$

Exercise. Perform the mean-field approximation for a homogeneous order parameter at the level of the continuous Ginzburg-Landau theory by adding a field term $-h\phi$. Discuss the phase transition in this model for $h = 0$ at $T = T_c$ and at $T < T_c$ and $h \to 0$.

As it turns out, the phase transition of the Ising model at $T = T_c$, $h = 0$ is a *second-order phase transition*, while for $T < T_c$, $h \to 0$, it is a *first-order phase transition*. This terminology derives from the singular (power-law) behaviour of the first and second derivatives of the free energy near the transition. At a first-order phase transition, there is a jump in the first derivative of the free energy or, in other words, the free energy has a kink. For a second-order phase transition, the free energy is continuous and differentiable, but at the transition its second derivative displays a discontinuity. We will encounter the distinction between first- and second-order transitions as well in Chapter 3, in the discussion of the thermal stability of DNA, and see that in this particular case, things are in fact more complicated.

Exercise. Draw the effective potential of the Ising model in Ginzburg-Landau theory with and without field. Draw the shape of the potential near the first and second-order transition to illustrate the above discussion.

Exercise. Calculate the dependence of the *susceptibility* on the distance of the critical point, $|T - T_c|$, i.e.

$$\chi = \partial_h \phi_0|_{h=0} . \tag{1.105}$$

Deduce the critical exponent of the susceptibility.

Exercise. Consider the phase transition from the disordered to the ordered state at $h = 0$ and at a finite value of h. Calculate the *specific heat*

$$C \equiv -T\partial_T F . \tag{1.106}$$

What do you notice? Determine the critical exponent.

Task. Are there also solutions to the Ginzburg-Landau equations which depend on space, e.g., one-dimensional profiles $\phi(x)$? What is their physical interpretation?

Phenomenological scaling. Having found the first examples of critical exponents, the next obvious question to ask is whether there are any relations between them. In order to answer this question we start from the homogeneous Ginzburg-Landau equation with field and express it in the form

$$m^2\phi + u\phi^3 = h . \tag{1.107}$$

The solutions to this equation describe a family of solutions $\phi = \phi(m^2, h)$. We have discussed before what happens if $h = 0$; if we approach the critical point at $T = T_c$, hence at $m^2 = 0$, we find the power-law dependence

$$\phi \sim h^{1/3}, \tag{1.108}$$

which yields another critical exponent.

Is there a way to combine the two control parameters m^2 and h? Let us introduce a scaled field variable

$$\psi = \phi h^{-1/3}. \tag{1.109}$$

Going back to Eq. (1.107) we find

$$m^2 h^{-2/3} \psi + u\psi^3 = 1. \tag{1.110}$$

Thus by redefining $x \equiv m^2 h^{-2/3}$ we obtain $x\psi + u\psi^3 = 1$ and the whole dependence on m^2 and h is now through a parameter combination. The general solution of the Ginzburg-Landau equation thus will have the *scaling form*

$$\phi(m^2, h) = h^{1/3} \psi(m^2 h^{-2/3}). \tag{1.111}$$

This equation states that the function $\phi h^{-1/3}$ has a *universal form*, and all curves parametrized by x will precisely collapse on this one universal curve.

What is the expression of this function? For the cubic polynomial, one can calculate it explicitly (take this as *Task*), but in most cases this is not feasible. It is more instructive to see the properties of this function in certain limits. We know that at $x = 0$

$$\psi(0) \sim u^{-1/3}. \tag{1.112}$$

For $x \to \infty$, the cubic term in Eq. (1.110) can be dropped and we find in that limit

$$\psi(x) \approx \frac{1}{x}, \tag{1.113}$$

hence

$$\phi \approx h^{1/3}(m^2 h^{-2/3})^{-1} = \frac{h}{m^2} \sim \frac{h}{T} \tag{1.114}$$

which is the *high-temperature* or *weak-field limit*; the final result is called *Curie's law* of paramagnetism in the context of magnetism.

For the opposite limit, $x \to -\infty$, we have $|x\psi| \gg 1$ and $|u\psi^3| \gg 1$ so that

$$\psi \approx \sqrt{|x|/u}\,. \tag{1.115}$$

We have seen this before: it is the *ordered phase* with $\phi \sim |m^2|^{1/2}$.

The idea to combine the control parameters into a single scaling variable has proved extremely fruitful for the theory of phase transitions. In the mid-sixties of the last century B. WIDOM proposed that quite generally the free energy per unit volume can be written in the scaling form

$$f(m^2, h) = |m^2|^{2-\alpha} f_s(h(m^2)^\Delta)\,, \tag{1.116}$$

where α is the *specific heat exponent*, and Δ the *gap exponent*. From this hypothesis, a whole sequence of critical exponent relations can be deduced that are now known to hold independent of spatial dimension (with some notable exceptions).

Task. From the above considerations of phenomenological scaling for the Ginzburg-Landau theory, deduce the relation between the critical exponents β, α, and Δ. Here, β is *not* to be confused with the abbreviation of $(k_B T)^{-1}$; it is the common notation for the exponent characterizing the temperature dependence of the order parameter via $m \sim |T - T_c|^\beta$.

Beyond Ginzburg-Landau theory. Ginzburg-Landau theory has permitted us to very easily find the equilibrium states ϕ_0 of the system by an almost trivial analytic calculation, the minimization of the effective potential $V_0(\phi)$, Eq. (1.100). Further, we have found the critical exponents. But we still have no clue about the validity of this approach. In contrast to our calculation of the partition function of the one-dimensional Ising model, spatial dimensions do nowhere appear explicitly, and the basic conflict between the one-dimensional result (no phase transition at $T > 0$) and the Ginzburg-Landau result (a phase transition at $T = T_c \neq 0$) persists. In order to finally resolve that conflict we now have to look for the effects of fluctuations around the mean-field solution. We expect that they will modify the behaviour we have obtained.

In order to compute the fluctuations around the mean-field solution we put $\phi = \phi_0 + \delta\phi$ and expand the Ginzburg-Landau Hamiltonian in $\delta\phi$. This leads to a partition function

$$Z = \exp[-\beta(H_0 + H_1)] \tag{1.117}$$

where βH_0 is the GL-value, and βH_1 is determined by a Gaussian integral over the $\delta\phi$-fluctuations which can be carried out exactly. It reads

$$\beta H_1 = \frac{1}{2} \int d^d\mathbf{x} \ln \left[-\nabla_\mathbf{x}^2 + m^2 + \frac{1}{2}\lambda\phi_0^2(\mathbf{x}) \right] \delta^d(\mathbf{x} - \mathbf{y})|_{\mathbf{x}=\mathbf{y}} . \qquad (1.118)$$

With the Fourier representation of the δ−function[18] we rewrite this as

$$\beta H_1 = \int d^d\mathbf{x} \int \frac{d^d\mathbf{k}}{(2\pi)^d} e^{-i\mathbf{k}\cdot\mathbf{x}} \ln \left[-\nabla_x^2 + m^2 + \frac{1}{2}\lambda\phi_0^2(x) \right] e^{i\mathbf{k}\cdot\mathbf{x}} \qquad (1.120)$$

where the term $\sim e^{-i\mathbf{k}\cdot\mathbf{y}}$ was first pulled through to the left and then put to $\mathbf{y} = \mathbf{x}$. If $\phi_0 = const.$ we find

$$\beta H_1 = \int \frac{d^d\mathbf{k}}{(2\pi)^d} \ln \left[\mathbf{k}^2 + m^2 + \frac{1}{2}\lambda\phi_0^2 \right] . \qquad (1.121)$$

Eq. (1.121) is the main result of this paragraph, and its consequences will now be discussed. Obviously, this integral is not well-behaved for large $\mathbf{k}$-values, indicating the breakdown of the continuum theory for small spatial scales.[19] We can render the integral finite by introducing a cut-off Λ, but already here it becomes evident that spatial dimension now enters in the calculation. In $d = 3$ we obtain

$$\beta H_1 = \frac{\lambda\phi_0^2\Lambda}{4\pi^2} - \frac{1}{6\pi} \left(m^2 + \frac{1}{2}\lambda\phi_0^2 \right)^{3/2} + \mathcal{O}(\Lambda^\alpha) \qquad (1.122)$$

where higher-order terms depending on Λ are summed up in the last term; further, terms which do not contain a dependence on the field have been dropped.

The expression (1.122) can be understood as an additional contribution to the effective potential in mean-field which we recall had the form

$$\beta V_0 = \frac{1}{2}m^2\phi_0^2 + \frac{\lambda}{4!}\phi_0^4 . \qquad (1.123)$$

---

[18]The Fourier representation of a function we use in the following is given by the integral

$$\widehat{f}(\mathbf{x}) = \int \frac{d^d\mathbf{k}}{(2\pi)^d} e^{i\mathbf{k}\cdot\mathbf{x}} f(\mathbf{k}) . \qquad (1.119)$$

Note that we frequently drop the $\widehat{}$ and distinguish between a function and its Fourier transform by its argument only.

[19]This phenomenon is called a *UV-divergence* in field theory since it occurs at large wavevectors, hence 'high energies'. We do not follow this point here in all its consequences, since this goes far beyond what is attempted here. The limited ambition here is to illustrate that a continuum theory has to be considered carefully: it is potentially dangerous to extrapolate its results down to microscopic scales. This may not show up in a purely mean-field approach, but if one wants to go beyond to include fluctuation effects, surprises can happen.

Expanding the root in βH_1 up to fourth order in ϕ_0 we obtain

$$\beta H_1 = -\frac{m^3}{6\pi}m^3 + \frac{\lambda\phi_0^2\Lambda}{4\pi^2} - \frac{\lambda m\phi_0^2}{8\pi} - \frac{1}{64\pi}\frac{\lambda^2}{m}\phi_0^4, \tag{1.124}$$

where the first term can be ignored since it does not depend on ϕ_0. We can now restore the original form of the GL-potential by the redefinitions

$$m_{eff}^2 = m^2 - \frac{\lambda\Lambda}{2\pi^2} + \frac{\lambda m}{4\pi} \tag{1.125}$$

and

$$\lambda_{eff} = \lambda + \frac{3}{8\pi}\frac{\lambda^2}{m}. \tag{1.126}$$

We stress again that, although the effective potential in $d = 3$ has the same form as the effective potential in mean-field, the new result does depend both on spatial dimension (the calculation is only valid in $d = 3$) and on the cutoff Λ we used in calculating the integral.

The vicinity of the transition: $m^2 \to 0$. We are now ready to take the final step and look at the effect of fluctuations at the critical point, or, more precisely, in the vicinity of the phase transition. In order to do this we have to look at the correlation function of the field $\phi(\mathbf{x})$ in the limit $m^2 \to 0$, since this will now give us additional information beyond the value of the order parameter itself, which we can compute from the fluctuation-modified effective potential.

Calling the *correlation function* of the field[20] $G(\mathbf{x})$, it fulfills the differential equation

$$(-\nabla_{\mathbf{x}}^2 + m^2)G(\mathbf{x}) = \delta^d(\mathbf{x}), \tag{1.127}$$

and, consequently, its Fourier transform is given by

$$G(\mathbf{k}) = \frac{1}{\mathbf{k}^2 + m^2}. \tag{1.128}$$

with, as before, $m \equiv \xi^{-1} \sim |T - T_c|^{1/2}$.

Now let's transform back to real space. We find

$$G(r) = \int \frac{d^d\mathbf{k}}{(2\pi)^d}\frac{e^{i\mathbf{k}\cdot\mathbf{x}}}{\mathbf{k}^2 + \xi^{-2}} = \xi^{2-d}\int \frac{d^d\mathbf{q}}{(2\pi)^d}\frac{e^{i\mathbf{q}\cdot\mathbf{x}/\xi}}{\mathbf{q}^2 + 1}, \tag{1.129}$$

---

[20]We made use of translational invariance of the system by shifting one of the arguments to $\mathbf{x} = \mathbf{0}$. In a translationally invariant system, the correlation function depends on the distance of field at the two selected points in space - we saw this already before in our computation of the correlation function in the one-dimensional spin chain.

where $\mathbf{q} = \xi\mathbf{k}$. The integrand has two asymptotic limits for $|x| = r$:

$$G(r) \sim \begin{cases} \frac{e^{-r/\xi}}{r^{(d-1)/2}}, & r \gg \xi \\[2mm] \frac{1}{r^{d-2}}, & r \ll \xi. \end{cases} \tag{1.130}$$

Note that the first expression indeed reduces to our previous result, Eq. (1.75), for $d = 1$. This means that, within Ginzburg-Landau theory, although there is a mean-field transition, fluctuations will destroy this transition and the original (exact) calculation is supported.

The existence of a second regime in which correlations decay algebraically signals the presence of a true ordered phase at a finite temperature, i.e., for $0 < T < T_c$, and we see that this is certainly possible for $d > 2$.

With this information we can now, finally, estimate the range of validity of the Ginzburg-Landau theory. Considering length scales on the order of the correlation length, we have

$$G(\xi) \sim \xi^{2-d} \sim |T - T_c|^{\frac{d-2}{2}}. \tag{1.131}$$

If we look at the ratio of G and the square of the mean-field value of the order parameter,

$$\frac{G(\xi)}{\phi_0^2} \sim |T - T_c|^{\frac{d-4}{2}}, \tag{1.132}$$

we see that, in the vicinity of the transition for $T \to T_c$, fluctuations grow indefinitely for $d < 4$. The dimension $d_u = 4$ is hence considered as an *upper critical dimension* above which mean-field theory becomes exact with respect to the critical exponents, i.e., fluctuations are negligible.[21] The dimension $d_l = 1$ is likewise a *lower critical dimension* at which the phase transition is destroyed by fluctuations.

This concludes our discussion of the Ising model. We leave this model at a point when, from the point of view of statistical physics, things become really interesting: how can we mathematically describe the transition in the range of dimensions $1 < d < 4$? In $d = 2$, the Ising model can be solved exactly by a transfer matrix approach (L. ONSAGER, 1944); in $d = 3$ the problem of the computation of the partition function was shown to be NP-complete

---

[21]In a more detailed calculation one can see that right at the upper critical dimensions, *logarithmic corrections* arise.

(S. ISTRAIL, 2000). It is here where renormalization-group methods need to be used, to which we will turn in a later stage of the book.

In the following, we will introduce three types of biomolecules that will play a role in the book: DNA, RNA and proteins. Subsequently we will see what aspects of their properties can be treated with the methods we learnt in this chapter.

Additional Notes

Equilibrium statistical mechanics is a well-developed theory on which many books have been and are being published, written to every taste of mathematical rigor. Four suggestions are the books by (K. HUANG, 2009), (M. PLISCHKE and B. BIRGERSEN, 2006), (L.E. REICHL, 2009) and (F. SCHWABL, 2010).

Books focussing on the renormalization group methods of statistical physics are, again as examples, the volumes by (L.P. KADANOFF, 2000) and (N. GOLDENFELD, 1992).

A highly recommendable introduction into biophysics, touching on many aspects of statistical physics, has been written by (P. NELSON, 2007). Two recent excellent biophysics books are by (H. SCHIESSEL, 2014), a book which is also based on statistical physics approaches, and further the broad introduction to the field for beginners by (F. CLERI, 2016).

References

F. Cleri
The Physics of Living Systems
Springer (2016)

N. Goldenfeld
Lectures on Phase Transitions and the Renormalization Group
Westview Press (1992)

K. Huang
Introduction to Statistical Physics
CRC Press (2009)

S. Istrail
Statistical Mechanics, Three-Dimensionality and NP-Completeness: I. Universality of Intractability of the Partition Functions of the Ising Model Across Non-Planar Lattices
Proc. 32nd ACM Symposium on the Theory of Computing (STOC00), ACM Press, 87-96 (2000)

E.T. Jaynes
Information theory and statistical mechanics I.
Phys. Rev. **106**, 620-630 (1957)

E.T. Jaynes
Information theory and statistical mechanics II.
Phys. Rev. **108**, 171-190 (1957)

L.P. Kadanoff
Statistical Physics. Statics, Dynamics and Renormalization
World Scientific (2000)

J. Majewski, H. Li, J. Ott
The Ising model in physics and statistical genetics
Am. J. Hum. Genetics **69**, 853-862 (2001)

P. Nelson
Biological Physics. Energy, Information, Life
Freeman (2007)

L. Onsager
Crystal statistics. I. A two-dimensional model with an order-disorder transition
Phys. Rev. **65**, 117-149 (1944)

M. Plischke and B. Birgersen
Equilibrium Statistical Physics
Scientific Publishers 3rd ed. (2006)

L.E. Reichl
A Modern Course in Statistical Physics
Wiley-Interscience 3rd ed. (2009)

H. Schiessel
Biophysics for Beginners
Pan Stanford Publishing (2014)

F. Schwabl
Statistical Mechanics
Springer (2010)

CHAPTER 2

Biomolecular Structure: DNA, RNA, Proteins

2.1   DNA, RNA AND PROTEINS: THE BUILDING BLOCKS

In this section we give a brief description of the basic chemical properties of the biomolecules whose statistical physical properties will occupy us in the following. The details given here are minimal; an indispensable reference for more detail is the book by B. ALBERTS et al. (2014).

DNA and RNA are charged polymers composed of three structural elements: a sugar, a phosphate group and the bases. The sugar gives the molecules their names: deoxyribose or ribose. Both are pentose rings with five carbon atoms; the arrangement of the carbons and their numbering is shown in Figure 2.1. The chemical formula of deoxyribose is $C_5H_{10}O_4$, of ribose $C_5H_{10}O_5$. The

Figure 2.1 The sugar ring of the backbone of both DNA and RNA. It contains an oxygen atom; the numbering of the carbon atoms is indicated.

Figure 2.2 A DNA single strand with a thymine base offering two hydrogen bonds to bind to a complementary base, adenine.

sugar builds, together with the phosphates, the backbone of DNA and RNA. The phosphates are attached to one sugar at the 5'-tail and to the next at the 3'-group, see Figure 2.2. Towards the opposite side, nucleotide bases are attached. Figure 2.3 displays the chemical structure of the four possible bases in DNA, two *pyrimidines*, and two *purines*, and the base uracil which replaces thymine in RNA. Neighbouring bases along a DNA strand experience *stacking interactions* when they are in registry, see Figure 2.4.

Consequently, a single-stranded DNA or RNA molecule is characterized by its base sequence and the orientation of the strand, e.g.

$$5' - ACTGTTTTACCCG - 3'.$$

The strand orientation is from the 5' to the 3' prime end; this is called the *sense strand*, while the direction 3' to 5' is the *antisense strand*. The bases can provide hydrogen bonds to a complementary base with which it can *hybridize* to form a double strand. This base pairing mechanism gives rise to the double-helical structure of the DNA molecule, see Figure 2.5.

In RNA, the base thymine (T) is replaced by uracil (U), and the sugar is ribose instead of 2-deoxyribose. In an organism RNA typically comes as a single strand, since it is the product of the *transcription* (i.e., the reading process) of a gene by the readout-molecule RNA polymerase. The basic variants of RNA are listed in Table 2.1.

Adenine (A)　　　Cytosine (C)　　　Guanine (G)　　　Thymine (T)
(in DNA)

Uracil (U)
(in RNA)

Figure 2.3 DNA and RNA bases. Uracil replaces thymine in RNA.

Figure 2.4 DNA and RNA stacking. Left: stacked bases, right: an unstacked base.

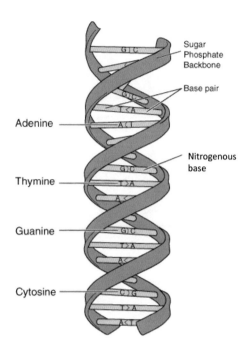

Sugar
Phosphate
Backbone

Base pair

Adenine

Nitrogenous
base

Thymine

Guanine

Cytosine

Figure 2.5 The DNA double helix. (From the *Talking Glossary of Genetic Terms*, US National Human Genome Research Institute.)

While DNA usually forms a double-strand between two complementary sequences, RNA frequently hybridizes with itself.[1] As a consequence of *self-hybridization*, an RNA base sequence can give rise to a rich *secondary structure*, composed from RNA single strands interspersed with helical, i.e., bound, regions.

These different structures correspond to the different functional roles the specific RNA molecule can play within its biological context. The most variable type is the *messenger RNA* (mRNA) which is synthesized as a transcript from the coding regions of DNA; it is the read-out of the transcription process by RNA polymerase. The schematic fold of a *transfer RNA* is illustrated in Figure 2.6. As its name says, transfer RNA has a transport function; it brings an mRNA transcript to the cellular bodies where protein synthesis occurs, the ribosomes. It has a characteristic cloverleaf structure, in which each of the leaves takes up a particular function.

---

[1]DNA can also self-hybridize, but this is the less frequent situation.

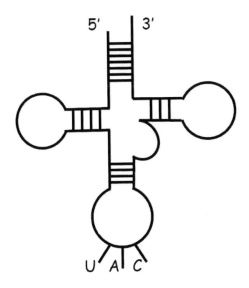

Figure 2.6 Transfer RNA in a schematic representation.

TABLE 2.1 RNA variants

RNA type	size (nt)	function
mRNA	variable	transcript: template for protein
tRNA	75-95	adapter molecule
rRNA	$\approx 10^3$	part of ribosome; protein synthesis

The original understanding of RNA as being a purely passive intermediate on the way from the genetic code of DNA to the functional properties of proteins has obviously evolved in the past years, in particular through the discovery of the mechanism of *RNA interference*, in which small non-coding *double-stranded* RNA molecules can regulate gene expression (A. FIRE et al., 1991). Apart from the classic variants, meanwhile a zoo of different RNA types has been found, some of which are indeed also double-stranded. RNA molecules are nowadays understood as being mostly non-coding, i.e., they do not help translate a DNA sequence into a protein. A list of the more recent examples is provided in Table 2.2. This list is only scratching the surface: look, e.g., at the list of RNA types provided by *Wikipedia*.

TABLE 2.2   Recent RNA variants

sRNA	small (non-coding) RNA, common name in bacteria, 75 − 400 bp
ncRNA	non-coding RNA, common name in eukaryotes
miRNA	microRNA, form a putative translational regulatory gene family
stRNA	small temporal RNA, often also called miRNA
snRNA	small nuclear RNA, includes eukaryote spliceosomal RNA
snmRNA	small non-mRNA, ncRNA
snoRNA	small nucleolar RNA, involved in rRNA modification
tmRNA	bacterial ncRNA, have both mRNA and tRNA function
siRNA	short (21-25 nt) interfering RNA, double-stranded, involved in gene silencing

One could be tempted to say that DNA is all *information*; by contrast, RNA is *information and structure*. DNA structure can vary between different conformational and topological states, but these are usually based on a double-stranded molecule. By contrast, the single-stranded RNA molecule can build a *sequence-dependent* three-dimensional structure, a *fold*, and in this way it acquires similar degrees of freedom as a protein. It is thus *not sufficient* to know the base sequence in order to characterize RNA: one also needs to know the fold in order to understand its function. We will turn to this issue in the next section.

Proteins are built up from twenty *amino acids*; they are listed in Table 2.3 (following A. VON HAESELER and D. LIEBERS, 2003). A triplet of DNA bases codes for an amino acid. Since the number of existing amino acids is lower than the combinatorial possibility based on the triplet rule, there is thus redundancy in the code. It appears typically, but not exclusively, in the last codon. Note that particular combinations of the bases also code for the start and stop of coding sequences.

Amino acids can further be distinguished by their *hydrophobicity*: hydrophobic amino acids will prefer to bury themselves inside of a protein fold in order to avoid water contact.

The sequence of amino acids is called the *protein primary structure*. Proteins also form *secondary structures* which are recurrent structural elements. These are the *α-helix*, and the *β-sheet*; as the name indicate they correspond to a helical and a folded, but fairly planar structure. Like DNA, the molecular building elements form hydrogen bonds with each other to stabilize these structures. On the next level of complexity of proteins arises the full fold, the

TABLE 2.3 Codons and amino acids; s.c.: starter codon

Bases			Name	Abbreviations	
A	G	AG	arginine	Arg	R
A	G	UC	serine	Ser	S
A	A	AG	lysine	Lys	K
A	A	UC	asparagine	Asn	N
A	C	UCAG	threonine	Thr	T
A	U	G	methionine*/s.c.	Met	M
A	U	UCA	isoleucine*	Ile	I
C	G	UCAG	arginine	Arg	R
C	A	AG	glutamine	Gln	Q
C	A	UC	histine	His	H
C	C	UCAG	proline*	Pro	P
C	U	UCAG	leucine	Leu	L
U	G	G	tryptophane*	Trp	W
U	G	A	stop codon		
U	G	UC	cysteine*	Cys	C
U	A	G	stop codon		
U	A	A	stop codon		
U	A	UC	tyrosine*	Tyr	Y
U	C	UCAG	serine	Ser	S
U	U	AG	leucine*	Leu	L
U	U	UC	phenylalanine*	Phe	F
G	G	UCAG	glycine*	Gly	G
G	A	AG	glutamic acid	Glu	E
G	A	UC	aspartic acid	Asp	D
G	C	UCAG	alanine*	Ala	A
G	U	G	starter codon (s.c.)		
G	U	UCA	valine*	Val	V

Note: Predominantly hydrophobic amino acids are marked with an asterisk; no distinction as to their degree of hydrophobicity is made.

Figure 2.7 Protein structure. (From the *Talking Glossary of Genetic Terms*, US National Human Genome Research Institute.)

tertiary structure. Finally, fully folded proteins can form *protein complexes* or *quaternary structures.* These elements are all summarized in Figure 2.7.

In the following, we will begin by our discussion with RNA structures, and use statistical mechanics methods to quantify them. A first step on this way is the graphical representation of RNA structures.

2.2 REPRESENTING RNA STRUCTURE

The basic structural elements that occur in RNA secondary structure are summarized in Figure 2.8. This collection already indicates the complexity one encounters when one wants to classify and predict these structures in the context of a complete molecule.

RNA structure is complex, but the number of the basic variable elements is only four, hence small when compared to the twenty amino acids of the proteins. It is useful to briefly list the ideas used to represent the self-hybridized configurations of an RNA molecule. Two kinds of representations are shown in Figure 2.9.

The top graph of Figure 2.9 shows an RNA molecule represented explicitly by vertices (the nucleotides) and full-line edges. The edges are the connections between the vertices and represent the hydrogen bonds, shown as dotted lines; in the following, we drop this distinction. But there is also, different from Figure 2.8, a second class of edges, drawn in broken lines.

Consider the difference between the two configurations, once without and once with the broken edges. Without the broken edges, the RNA could be cut in two without affecting the hybridized chains. This is not the case anymore when the broken edges are present: the RNA structure now has a *pseudoknot*, intertwining separate paired regions along the chain.

This property becomes clear in the bottom graph. It can be thought to be obtained from the top graph by 'pulling at the ends' of the RNA chain. The *arc* or *rainbow diagram* clearly shows the 'overlapping' bonds for the pseudoknot.

There are formal ways to represent these RNA configurations which can be used in computer implementations (see, e.g., G. VERNIZZI et al., 2004). The first is a bracketing rule: an unpaired base is given by a dot, while a paired base is described by a bracket, indicating opening or closing of a bond. The

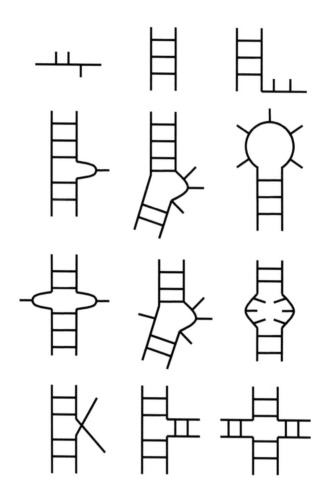

Figure 2.8 Schematic representation of the elements of RNA secondary structure; backbone and base pairs are both indicated as black bars, so no difference is made between covalent and hydrogen bonds. From top left to bottom right: single strand; duplex; duplex with dangling end; single-nucleotide bulge; three-nucleotide bulge; hairpin: stem (duplex region) and loop; mismatch pair or symmetrical loop; asymmetric internal loop; symmetric internal loop; two-stem junction; three-stem junction; four-stem junction.

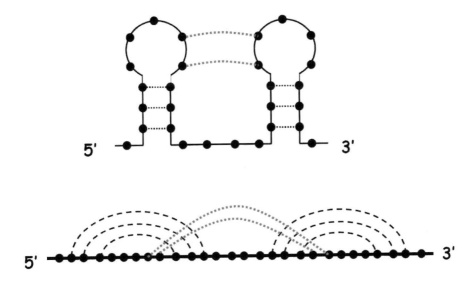

Figure 2.9 (**See color insert.**) Graphical representation of RNA with pseudoknots (see text).

top graph in Figure 2.9 would thus be represented as

$$.(((.....)))....(((.....))).\tag{2.1}$$

for the case without pseudoknot, and as

$$.(((...[)))....((((]...))).\tag{2.2}$$

when the pseudoknot bonds are present. Each additional pseudoknot requires a new bracket-type to be introduced.

Alternatively, one can represent the bonds along the chain also in an array or matrix form. For this one writes the sequence of L nucleotides as an $L \times L$ *contact matrix* C with elements $C_{ij} = 1$, if i is paired to j, or zero otherwise. One can also interpret the pairing between any two bases i and j as a transposition of the elements $\{i, j\}$ and associate a permutation structure to it via $\sigma(i) = j$ if i, j are paired, and $\sigma(i) = i$ if not. For the example sequence $\{5' - CUUCAUCAGGAAAUGAC - 3'\}$ we give the pseudo-knotted structure in dot-bracket notation in the first row and the permutation structure in

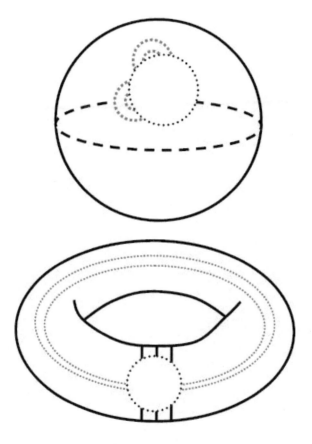

Figure 2.10 (**See color insert.**) Drawing RNA on a sphere and on a torus.

the second and third:

$$\sigma \; = \tag{2.3}$$

$$\begin{pmatrix} . & (& (& (& . & [& [& [&) &) &) & . & . &] &] &] & . \\ 1 & 2 & 3 & 4 & 5 & 6 & 7 & 8 & 9 & 10 & 11 & 12 & 13 & 14 & 15 & 16 & 17 \\ 1 & 11 & 10 & 9 & 5 & 16 & 15 & 14 & 4 & 3 & 2 & 12 & 13 & 8 & 7 & 6 & 17 \end{pmatrix}$$

This is an involution since σ^2 is the identity permutation.

Finally, there is an even more fundamental way to look at RNA *pseudoknots*. If one closes the endpoints of a *rainbow diagram* to let it form a closed loop, one obtains a *circle diagram*. In this diagram, RNA structures with and without pseudoknots are distinguished by the presence or absence of edge crossings, when all of them are drawn either inside or outside the circle diagram.

Exercise. Draw the circle diagram for the kissing hairpin, with and without the pseudoknot bonds, and with bonds lying either all inside or outside the circle.

It is instructive to draw the circle diagram on the surface of a sphere, with the edges outside of the circle. Figure 2.10 (top) shows this for a non-pseudoknotted case. By contrast, a pseudoknotted configuration can be drawn without edge crossings on the surface of a torus (Figure 2.10, bottom).

The relationship between RNA structures and topology can be made quantitative through the *Euler characteristic* of the surface on which they can be drawn without crossing the bonds. The Euler characteristic is defined as

$$\chi \equiv V - E + F \tag{2.4}$$

where V are the vertices (nucleotides), E the edges (bonds), and F the faces of closed loops. RNA chains without pseudoknots have $\chi = 1$. For the kissing hairpin, one has $\chi = -1$. This result can also be expressed in terms of the *genus* g of the surface, which is $\chi = 1 - 2g$ for the case at hand. It is then clear that the kissing hairpin can be represented by a torus, for which $g = 1$. The notion of topology in RNA folds will be discussed further in the following section.

2.3 COMPUTING RNA SECONDARY STRUCTURE: COMBINATORICS

We have seen why the knowledge of RNA secondary structure is important - for information processing in biology and biotechnology. In this section we want to learn how to predict RNA structure. Predicting RNA structure, as we will see, has three aspects. The first aspect is combinatorial: how can we classify and compute the *possible* structures?

Maximal base pairing. We begin the discussion by formalizing the graphical representations we have just introduced.

If the primary structure of an RNA molecule - i.e., its sequence - of length n is denoted by a string

$$r = r_1...r_n, \tag{2.5}$$

its secondary structure can be described as a set S of *disjoint pairs* (r_i, r_j) for $1 \leq i < j \leq n$. Considering the bases as vertices on a graph, and all pairings as the edges of that graph, the secondary structure is a matching in a graph $G = (V, E)$ with the properties that V contains a vertex for every base pair r_i, $i = 1, ..., n$, and E contains an edge (u, v) if and only if $u, v \in V$ are complementary bases.

In the prediction of secondary structure, we want to first exclude *pseudoknots*. In the above notation such a knot occurs when a base r_i is paired with a base r_j, and a base r_k with r_l such that $i < k < j < l$: the pairs overlap. It is this condition which we reject for the moment.

The prescription we have given allows to cover all possible secondary structures, except for the occurrence of pseudoknots. However, we would not know which one to select from the set of structures. A natural selection principle is based on free energy. Assuming that we can associate with every complementary base pair an energy contribution $\alpha(r_i, r_j) = -1$ and put $\alpha = 0$ otherwise, we see that the minimal (free) energy $E(S) = \sum_{(r_i, r_j)} \alpha(r_i, r_j)$ will be attained for the RNA structure with the maximal number of base pairs.

In this simplified version which treats all bound pairs on equal footing, the problem of RNA secondary structure determination can be solved by a *dynamic programming* approach. With the assumption that $\alpha(r_i, r_j)$ is independent of all other pairs and positions of r_i and r_j in the sequence, we can compute the free energy of a substring $r_i...r_j$ of the RNA molecule disregarding the surrounding $r_1...r_{i-1}$ and $r_j + 1...r_n$. In this way, we can use solutions for smaller strings to obtain those for larger strings and obtain a recursive solution.

The resulting dynamic programming algorithm is easy to develop; the possibilities that can arise in building up the structure are shown in Figure 2.11. There are three alternatives:

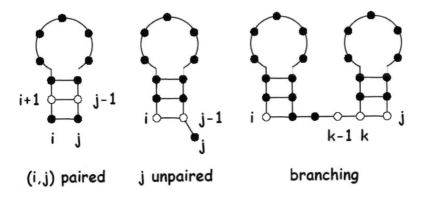

(i,j) paired   j unpaired   branching

Figure 2.11 Dynamic programming recursion for RNA structure (see text).

- Add paired bases to an optimal structure for the subsequence $i+1, j-1$;
- Add an unpaired base to an optimal structure for the subsequence $i, j-1$; this possibility arises symmetrically for the other end of the molecule;
- Combine two optimal substructures $i, k-1$ and k, j.

The latter step obviously runs in problems when pseudoknots are to be accounted for.

The resulting recursion can be summarized as

$$
E(S_{i,j}) = \min
\begin{cases}
E(S_{i+1,j-1}) - 1 \\[2ex]
E(S_{i+1,j}) \\[2ex]
E(S_{i,j-1}) \\[2ex]
\min\{E(S_{i,k-1}) + E(S_{k,j})\}, \quad i < k \leq j
\end{cases}
\tag{2.6}
$$

In order to initialize the algorithm, one requires $E(S_{i,i}) = 0$ for $i = 1, .., n$ and $E(S_{i,i-1}) = 0$ for $i = 2, .., n$.

Let us illustrate this scheme with a small example (S. R. EDDY, 2004). We take the RNA sequence $r = GGGAAAUCC$; for each allowed pair AU, GC we take $\alpha = -1$, and $\alpha = 0$ otherwise. The pairing matrix $(r \times r)$ at the initialisation step reads

$$
\begin{bmatrix}
  & G & G & G & A & A & A & U & C & C \\
G & 0 & & & & & & & & \\
G & 0 & 0 & & & & & & & \\
G & x & 0 & 0 & & & & & & \\
A & x & x & 0 & 0 & & & & & \\
A & x & x & x & 0 & 0 & & & & \\
A & x & x & x & x & 0 & 0 & & & \\
U & x & x & x & x & x & 0 & 0 & & \\
C & x & x & x & x & x & x & 0 & 0 & \\
C & x & x & x & x & x & x & x & 0 & 0
\end{bmatrix}
\tag{2.7}
$$

where the lower half (x) is not used, and the diagonal and its left neighbour are put to 0.

In the computation, one finds that entry Y depends on all entries denoted by y, and $w = (i+1, j)$,

$$
\begin{bmatrix}
  & G & G & G & A & A & A & U & C & C \\
G & 0 & & & & & & & & \\
G & 0 & 0 & & & & & & & \\
G & x & 0 & 0 & y & y & y & y & Y & \\
A & x & x & 0 & 0 & & & w & y & \\
A & x & x & x & 0 & 0 & & & y & \\
A & x & x & x & x & 0 & 0 & & y & \\
U & x & x & x & x & x & 0 & 0 & y & \\
C & x & x & x & x & x & x & 0 & 0 & \\
C & x & x & x & x & x & x & x & 0 & 0
\end{bmatrix}
\tag{2.8}
$$

so that one obtains the following result by going through the recursion

$$
\begin{bmatrix}
  & G & G & G & A & A & A & U & C & C \\
G & 0 & 0 & 0 & 0 & 0 & 0 & -1 & -2 & (-3) \\
G & 0 & 0 & 0 & 0 & 0 & 0 & -1 & -2 & (-3) \\
G & x & 0 & 0 & 0 & 0 & 0 & -1 & (-2) & -2 \\
A & x & x & 0 & 0 & 0 & 0 & (-1) & -1 & -1 \\
A & x & x & x & 0 & 0 & (0) & -1 & -1 & -1 \\
A & x & x & x & x & 0 & (0) & -1 & -1 & -1 \\
U & x & x & x & x & x & 0 & 0 & 0 & 0 \\
C & x & x & x & x & x & x & 0 & 0 & 0 \\
C & x & x & x & x & x & x & x & 0 & 0
\end{bmatrix}
\tag{2.9}
$$

The minimum structure can be found by tracing back through the table following a diagonal trace (i, j) to $(i+1, j-1)$. The result is shown in Figure 2.12; the trace back is indicated by terms in brackets. The complexity of the algorithm to compute the structure is $O(n^3)$, since there are n^2 entries, and

Figure 2.12 Exemplary RNA structure for the sequence $GGGAAAUCC$ as obtained from the simple algorithm.

each computation is $O(n)$. The trace back to find the structure takes again linear time $O(n)$, if back pointers are used.

Following the original approach by R. NUSSINOV and A. B. JACOBSON, 1980, there have been several further developments of algorithms for *RNA secondary structure prediction*. M. ZUKER, 1989, made an important contribution by developing a tool for the determination of sub-optimal folds. More recent developments are targeted towards the inclusion of pseudoknots; we comment on them in the Additional Notes.

The basic approach to RNA secondary structure, as we have seen, is to compute the possible combinations by a recursion, and to score them by a suitably defined free energy. Thinking back to what we saw for DNA, one might question whether the first part - the computation of the configurations - cannot also be done based on our main tool from statistical mechanics, the partition function Z. This is what follows.

2.4   THE RNA PARTITION FUNCTION

The partition function of a chain of L nucleotides can be written in the following very general way as (H. ORLAND and A. ZEE, 2002)

$$\mathcal{Z} = \int \prod_{k=1}^{L} d^3\mathbf{r}_k \, \mathcal{F}(\{\mathbf{r}\}) \, Z_L(\{\mathbf{r}\}). \tag{2.10}$$

Here, $\mathbf{r}_k$ is the position vector of the k-th nucleotide. The function

$$\mathcal{F}(\{\mathbf{r}\}) = \prod_{i=1}^{L-1} f(|\mathbf{r}_{i+1} - \mathbf{r}_i|) \tag{2.11}$$

is a model-dependent function of the molecular geometry, and takes into account the steric constraints of the chain. Standard choices for f are

$$f(r) = \delta(r - \ell), \tag{2.12}$$

if the nucleotides are connected by rigid rods of size ℓ, or

$$f(r) = \exp[-(r - \ell)^2/6\eta^2] \tag{2.13}$$

if the rods are taken as elastic springs with η as a measure of their stiffness.

The partial partition function Z_L in Eq. (2.10) counts the different configurations of paired bases. It can be defined by the series

$$Z_L(\{\mathbf{r}\}) = 1 + \sum_{(ij)} V_{ij}(\mathbf{r}_{ij}) + \sum_{(ijkl)} (V_{ij}(\mathbf{r}_{ij})V_{kl}(\mathbf{r}_{kl}) + V_{ik}(\mathbf{r}_{ik})V_{jl}(\mathbf{r}_{jl})) + \dots \tag{2.14}$$

with $\mathbf{r}_{ij} \equiv |\mathbf{r}_i - \mathbf{r}_j|$, and where the summation index (ij) denotes all pairs with $i < j$, $(ijkl)$ all quadruplets with $i < j < k < l$, and so forth. The first term of the series describes the binding energy between the bonds i and j, with the indices running all along the chain, the second the binding of i and j in a configuration together with k and l, and so forth. The factors

$$V_{ij} \equiv \exp(-\beta\varepsilon_{ij}v_{ij}(\mathbf{r}_{ij}))\theta(|i - j| > 4) \tag{2.15}$$

are the Boltzmann factors associated with a (4×4)-dimensional symmetric matrix ε_{ij} of bond energies between the i-th and j-th bases at a distance $\mathbf{r}_{ij}$; $\beta = 1/(k_BT)$ is the inverse thermal energy as before. In Eq. (2.15), the factor $v_{ij} = v(|\mathbf{r}_i - \mathbf{r}_j|)$ is a short-range attractive interaction between the bases. The Heaviside function $\theta(|i - j| > 4)$ expresses the sterical constraint which prohibits hybridization of the bases in closest proximity to each other. Finally, note that $V_{ii} = 0$.

The series (2.14) can be expressed in integral form by introducing a set of $i = 1, ..., L$ *Hermitian matrices*[2] φ_i, $i = 1, ..., L$, using the expression for the

---

[2]A square matrix is hermitian if it is *self-adjoint*, i.e., fulfills $A_{ij} = \overline{A}_{ji}$, where the overbar stands for the complex conjugate. We encountered examples of *self-adjoint matrices* in Chapter 1 in the discussion of the Ising model, e.g., the matrix σ_3.

ordered matrix product

$$\mathcal{O}_\pi[\{\varphi\}] \equiv \prod_{l=1}^{L}(1 + \varphi_l) \equiv (1 + \varphi_1)(1 + \varphi_2) \cdots (1 + \varphi_L). \tag{2.16}$$

The matrices have the dimension $(N \times N)$. The parameter N is introduced on purely formal grounds; as will be seen below, it allows to organize the expansion of the partition function according to the topology of RNA. It can, however, also be related to physical quantity, the concentration of a chemical which favors pseudoknots. Such chemicals, e.g., are divalent ions such as Mg^{2+}, and the relationship is given via the identification $N^{-2} \equiv \exp \beta \mu$ where μ is the chemical potential of the ions.

Coming back to the calculation, the resulting formula for $Z_L(N)$ is

$$Z_L(N) = A_L(N)^{-1} \int \prod_{k=1}^{L} d\varphi_k e^{-\frac{N}{2}\sum_{ij}((V^{-1})_{ij}Tr(\varphi_i\varphi_j))} \frac{1}{N} Tr\mathcal{O}_\pi[\{\varphi\}] \tag{2.17}$$

where the normalization factor is given by

$$A_L(N) = \int \prod_{k=1}^{L} d\varphi_k e^{-\frac{N}{2}\sum_{ij}((V^{-1})_{ij}Tr(\varphi_i\varphi_j))}. \tag{2.18}$$

In both expressions, V is an $(L \times L)$-matrix with entries V_{ij}.

After the introduction of the matrices, the partition function given by Eq. (2.17) looks like a Gaussian integral in the matrices φ_k over a product observable in the φ_k. The product of the terms $(1 + \varphi_l)$ evaluates into a polynomial of order L and hence we have to perform Gaussian integrals over all the contributions of this polynomial from order 1 to L - thus nothing but the moments of the partition function.

The important thing is now that the introduction of the matrix-dimension N gives a handle to reorganize the series representation. One can show that the result is an asymptotic series in $1/N$ of the form

$$Z_L(N) = 1 + \sum_{i<j} V_{ij} + \sum_{i<j<k<l} V_{ij}V_{kl} + \dots + \frac{1}{N^2} \sum_{i<j<k<l} V_{ik}V_{jl} + \dots \tag{2.19}$$

The comparison of the two expressions eqs.(2.14), (2.19) shows that both coincide for $N = 1$; the latter, however, for $N > 1$ now contains information about the *topology* or RNA. The $O(1)$-terms of Eq. (2.19) yield the planar secondary structures of RNA, while the $1/N^2$-terms correspond to RNA tertiary structure.

The secondary structure can be evaluated within this formalism by approximating the integral (2.17) by its saddle-point value in the limit $N \to \infty$. In order to be able to perform this limit, the integral has to be transformed by a Hubbard-Stratonovich transform into an expression in which the dependence on the parameter N - which is still the dimension of the matrices - becomes explicit. How such a transform is set up can be found in Chapter 1; it is left an as *Task* for the readers to apply it to the present case. Those who want to see it explicitly can find it in the paper by (H. ORLAND and A. ZEE, 2002), which also discusses how Eq. (2.19) is obtained.

The result of the calculation is the expression (with C as an irrelevant normalization factor)

$$Z_L(N) = \frac{1}{C} \int dA e^{-\frac{N}{2}Tr\, A^2 + NTr\, \ln M(A))} M^{-1}(A)_{L+1,1} \qquad (2.20)$$

where the integral runs over all Hermitian matrices A of dimension $(L+1) \times (L+1)$. M is a matrix function of A given by

$$M_{ij} = \delta_{ij} - \delta_{i,j+1} + i\sqrt{V_{i-1,j}}A_{i-1,j} \qquad (2.21)$$

and the symbol for trace, Tr, means - as in Chapter 1 - the sum over the diagonal elements.

The saddle-point of Eq. (2.20) follows from the variation $\delta S(A)/\delta A = 0$, where S is

$$S(A) \equiv \frac{1}{2}Tr\, A^2 - Tr\, \ln M(A). \qquad (2.22)$$

The stationary point is given by

$$A_{lk}^0 = i\sqrt{V_{lk}}(M^{-1})_{l,k+1}. \qquad (2.23)$$

Introducing $G_{ij} \equiv (M^{-1})_{i+1,j}$, and using the identity $\sum_j M_{ij}(M^{-1})_{jk} = \delta_{ik}$ one obtains the so-called *Hartree equation*

$$G_{i+1,k} = \delta_{i+2,k} + G_{ik} + \sum_j V_{i+1,j}G_{i,j+1}G_{j-1,k}. \qquad (2.24)$$

The Hartree equation is a recursion relation which can be solved for the boundary condition $G_{i,i+l} = 0$ for $l \geq 0$. Then, G_{ij} is the partition function of the secondary structure of a chain starting at base j and ending at base i, just as it is used in dynamic programming algorithms. We have therefore, as the saddle point of the partition function, recovered a formal expression which allows to compute RNA secondary structure. We can use the result Eq. (2.24) to compute, recursively, the structure of RNA; note that the energy evaluation is contained in V.

From this result, we can now branch off into two directions: we apply Eq. (2.24), or we go on to evaluate the next terms in the series. This would amount to go for pseudoknots. We will do both, and begin with the pseudoknots.

Pseudoknots. The computation of pseudoknots based on the theory described in the previous section is rather involved, since the computation of the higher order terms in $1/N^2$ requires the use of field-theoretic methods which are out of the scope of this book.[3] The interested reader is asked to consult the original literature at this point.

Here, in what follows, we want restrict ourselves to obtain a general idea of the occurrence of pseudoknots, based on the topological theory. The basic question we address is: as the RNA structures become more and more complex, how many pseudoknots will arise? This theory allows us to readily determine the *number of RNA pseudoknots* according to their topological character, and to see how it evolves as the sequence increases in length. For this calculation, some simplifying assumptions can be made. If any possible pairing between nucleotides is allowed (independent of the identity of the base pair and its location along the chain), and if all of these pairings will occur with equal probability, then the matrix V_{ij} has identical entries, which we suppose as $v > 0$ for all i, j.[4]

The computation of the integral $Z_L(N)$ in this simplified case runs as follows. The original expression for $Z_L(N)$, Eq. (2.19), can be rewritten using the integral transforms introduced before as

$$Z_L(N) = A^{-1}(N) \int d\sigma e^{-\frac{N}{2v}Tr\sigma^2} \frac{1}{N} Tr(1+\sigma)^L, \qquad (2.25)$$

where σ is a single $(N \times N)$-matrix. The normalization factor in this case is explicitly given by

$$A(N) = \int d\sigma e^{-\frac{N}{2v}Tr\sigma^2} = \left(\frac{\pi v}{N}\right)^{\frac{N^2}{2}} 2^{N/2}. \qquad (2.26)$$

---

[3]We were pushing the limits a bit already. The method used here is well known as a $1/N$-expansion in the context of field theory in statistical mechanics and the theory of elementary particles, and may be a bit difficult to digest for a bioinformatician. Note, however, that it is actually equivalent to approaches from a completely different context in computer science, the *context-free grammars* (see Additional Notes).

[4]A technical detail: in order for the computations to make sense when we do this, we need to make V_{ij} positive definite. We can do this by adding an arbitrary real number a to the diagonal elements. Since no diagonal terms appear in the original series ($V_{ii} = 0$), this number plays only a formal regularizing role, and the final result can be shown to be independent of its choice.

In order to solve the Gaussian matrix integral (2.25) it is convenient to introduce the *spectral density* ϱ_N of the matrix σ,

$$\varrho_N(\lambda) \equiv A^{-1}(N) \int d\sigma e^{-\frac{N}{2v}Tr\sigma^2} \frac{1}{N} Tr\delta(\lambda - \sigma). \qquad (2.27)$$

This is a convenient trick - in the same spirit as the Hubbard-Stratonovich transformation we used for the Ising model in Chapter 1. With the spectral density we can represent $Z_L(N)$ as

$$Z_L(N) = \int_{-\infty}^{+\infty} d\lambda \varrho_N(\lambda)(1 + \lambda)^N, \qquad (2.28)$$

where the identity $\int_{-\infty}^{+\infty} d\lambda\, \varrho_N(\lambda) = 1$ was used in Eq. (2.25). Hence, we have reduced a multidimensional problem to a one-dimensional integral.

One can now introduce the (exponential) generating function of $Z_L(N)$,

$$G(t, N) \equiv \sum_{L=0}^{\infty} Z_L(N)\frac{t^L}{L!} = \int_{-\infty}^{+\infty} d\lambda \varrho_N(\lambda)e^{t(1+\lambda)}, \qquad (2.29)$$

and we need the explicit form of $\varrho_N(\lambda)$. The latter is known from *Random Matrix Theory* (see M. L. MEHTA, 1991). It can be expressed in terms of a series of *Hermite polynomials* $H_k = (-1)^k \exp(x^2)(d^k/dx^k) \exp(-x^2)$,

$$\varrho_N(\lambda) = \frac{e^{-N\lambda^2/(2v)}}{\sqrt{2\pi vN}} \sum_{k=0}^{N-1} \binom{N}{k+1} \frac{H_{2k}(\lambda\sqrt{N/2v})}{2^k k!}. \qquad (2.30)$$

With this one obtains for $G(t, N)$ the expression

$$G(t, N) = e^{\frac{vt^2}{2N}+t} \frac{1}{N} L_{N-1}^{(1)}\left(-\frac{vt^2}{N}\right), \qquad (2.31)$$

where $L_N^{(1)}(z)$ are the generalized *Laguerre polynomials*.[5] The series expansion of $G(N, t)$ in t now gives the first coefficients of $Z_L(N)$. Putting $v = 1$, one can write

$$Z_L(N) = \sum_{L=0}^{\infty} \frac{a_{L,g}}{N^{2g}} \qquad (2.32)$$

where the coefficients $a_{L,g}$ determine the *number of diagrams at fixed length* L and *fixed genus g*.

---

[5]The generalized Laguerre polynomials are defined by

$$L - n^{\alpha}(x) = \frac{(\alpha + 1)_n}{n!} [_1F_1(-n; \alpha + 1; x)]$$

where $(\alpha)_n = \Gamma(x + n)/\Gamma(n)$ is the *Pochhammer symbol* and $[_1F_1(a; b; z)] = \sum_{k=0}^{\infty} \frac{(a)_k}{(b)_k} \frac{z^k}{k!}$ is the *confluent hypergeometric function of the first kind*.

Now that we have mastered this excursion into the functions of classical mathematical physics, we want to see what we can learn from this result. How can we interpret it?

Let's look at an example. For $L = 8$, we have

$$
\begin{aligned}
Z_8(N) \;=\; & 1 + 28v + 140v^2 + 140v^3 + 14v^4 + \qquad\qquad (2.33)\\
& + \frac{1}{N^2}(70v^2 + 280v^3 + 70v^4) + 21\frac{v^4}{N^4}
\end{aligned}
$$

The meaning of the numbers is:

- The power of v is the number of nucleotides that are paired;

- The power of N^{-2} is the genus g of the structure (i.e., the *genus* of the surface on which it can be drawn without bond crossings);

- Putting $v = 1$, one obtains the number of structures for a given genus g;

- Putting $N = 1$, one obtains the number of structures with a given number of bonds.

Thus, there are 28 planar structures with one bonded pair, 70 structures with two bonded pairs on a torus, etc. The number of structures with genus $g = 1$ is 420, while the number of structures with four pairs (the maximal number) is 105.

After this example, we finish by reading off some general characteristics from formula (2.32). An analysis of this series for a given length $L \gg 1$ shows that the normalized number of pseudoknots $a_{L,g}/Z_L(1)$ is always peaked at a characteristic genus of roughly $g_c(L) \sim 0.23L$. For fixed L, the maximally achievable genus comes out to lie at $g \leq L/4$. More interestingly, in order for a structure to have a given genus g, it needs to have a length $L \geq 4g$, i.e., for an RNA structure with $g = 7$, the sequence must be at least 28 bases long.

2.5   PROTEIN FOLDING AND DOCKING

The fact that there are twenty building blocks for a protein - the amino acids - makes the problem of relating sequence to structure even more difficult to solve than for RNA with its four base units, the nucleotides.

The puzzling feature of the *protein folding problem* is best illustrated by invoking a famous paradox, first stated by (C. LEVINTHAL, 1969). If one starts out with the protein as a linear polymer and wants to fold it into its 3-D spatial structure which one assumes known from X-ray crystallography, each residue in the chain has about ten times more conformational positions available as in its native state. Thus the total number of conformational shapes for a 100-residue protein is 10^{100}. Supposing a conformational change at the single residue level happens at a picosecond rate, a fully random exploration of the landscape would take 10^{81} years to find the 'right' structure.

Obviously, this is not what happens. Following a metaphor coined by P. WOLYNES, 2001, the random walk of the protein conformations in Levinthal energy landscape resembles that of a golf course:[6] all energy levels are equal, except for the right one.

The consensus on protein folding that has emerged up to now can be summarized as follows:

- 'Small' proteins are two-state folders. This problem then is similar to that of DNA denaturation and hybridization which we will consider in the next chapter.

- 'Medium' proteins are 'complex' because *many* details of the composition matter for the final folded state;

- 'Large' proteins do not have the problem we discuss: they simply do not fold spontaneously. They are folded with the help of a specialized molecular machinery, the *chaperones*. We will not address this topic here any further.

We will here only consider the 'medium' proteins.

Medium proteins. It is more reasonable to think of a characterization of the folding process as an ensemble of possible pathways which can be visualized as trajectories in a complex energy landscape. The energy space of protein conformations is what is sometimes called a 'rugged' landscape, with many minima of near energetic degeneracy.

An elegant concept to characterize this landscape is the *funnel*, shown in Figure 2.13 (H.S. CHAN and K.A. DILL, 1998). In this plot, free energy is plotted versus conformation. The folding process of a protein corresponds to a descent in the funnel. In the beginning, near the top of the funnel, a large number of possible pathways exist for the protein. The more it descends in the funnel, the more limited becomes the number of available pathways: the funnel narrows down towards the conformation with the minimal free energy.

---

[6]This should hold at least for the putting green.

(a)   **(b)**

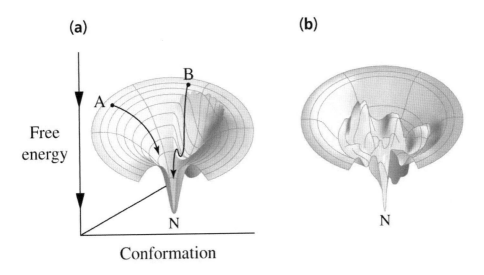

Free
energy

Conformation

Figure 2.13 (**See color insert.**) The folding funnel; a) folding paths from two different initial conditions: A follows a direct path whereas B is trapped in a metastable intermediate; b) rugged energy landscape with multiple minima at intermediate energies. Note that the 'true' equilibrium has a much lower free energy. [Reprinted with permission from John Wiley & Sons (H.S. CHAN and K.A. DILL, 1998).]

Let's try to quantify this idea a bit. For this we turn to a basic model from statistical physics that rationalizes the idea of a random energy landscape. The *Random Energy Model* (REM) was invented by B. DERRIDA (1980) with the idea to represent a class of models of disordered systems with many nearly degenerate minima.[7] For its application to protein folding, we follow J.N. ONUCHIC et al. (1997) and consider a simple model with two basic variables, the energy[8] E, and one additional quantity, which we take as the

---

[7]The most prominent example of such systems is the '*spin glass*', which is a disordered ferromagnetic system in which a large number of ferromagnetic couplings have been replaced by antiferromagnetic couplings. Since the latter favor the orientation of the neighboring magnetic moments in antiparallel direction, the system can become 'frustrated' since the spins may underlie conflicting conditions for the orientation of their magnetic moments. This gives rise to a large degeneracy of the microstates of the system.

[8]Which we assume as averaged over the solvent.

fraction of *native-like contacts*, Q, hence a measure for the structural similarity between a given protein conformation and the native one. For the native structure, we take $Q = 1$. Q serves as our conformational coordinate.

For this model we can define a number of useful physical quantities like the thermal average of the energy, $\overline{E}(Q)$, the roughness of the energy landscape (i.e., the fluctuations in energy), $\sqrt{\Delta E^2(Q)}$, and the density of states, $\Omega(E, Q)$. The entropy of a configuration, e.g., is then given by $S(E, Q) = k_B \ln \Omega$.

One key feature of the REM-model is the assumption of a Gaussian distribution of energy states,

$$P(Q, E) = \frac{1}{\sqrt{2\pi \Delta E^2(Q)}} \exp\left(-\frac{(E - \overline{E}(Q))^2}{2\Delta E^2(Q)}\right). \tag{2.34}$$

If a protein has N residues, the total number of its conformations is

$$\Omega_0 = \gamma^N \tag{2.35}$$

where γ is the number of configurations per residue. This number is amenable to simplifications: e.g., when only the backbone coordinates are taken, $\gamma \approx 5$, while the inclusion of excluded-volume effects allows $\gamma \approx 1.5$.

Since by going down the energy funnel the folded structures more and more resemble the native state, the total number of configurations decreases: the native structure has a unique backbone conformation. If one calls $\Omega_0(Q)$ the density of conformational states of measure Q and $S_0(Q)$ the corresponding entropy, then the density of conformations with associated energy E is given by

$$\Omega(Q, E) = \Omega_0 P(Q, E) \tag{2.36}$$

and the total entropy is

$$S(Q, E) = S_0(Q) - k_B \frac{(E - \overline{E}(Q))^2}{2\Delta E^2(Q)}. \tag{2.37}$$

At thermal equilibrium, the most probable energy is given by the maximum of the distribution

$$E_{mp} = \overline{E}(Q) - \frac{\Delta E^2(Q)}{k_B T}, \tag{2.38}$$

and the corresponding values of Ω and S at this maximum are easy to compute using eqs.(2.36) and (2.37). Take this as an *Exercise*.

Further, we can compute the free energy of a misfolded structure of a given structural similarity Q, and at a given temperature T

$$F(Q,T) = E_{mp} - TS(E_{mp}, Q) = \overline{E}(Q) - \frac{\Delta E^2(Q)}{2k_B T} - TS_0(Q). \qquad (2.39)$$

Without precise knowledge about the dependence of this expression on Q, further analysis is not possible. To proceed we consider the simplest case, namely that the free energy has two minima, one at $Q \approx 0$, corresponding to an *ensemble* of collapsed misfolded states with a varying degree of ordering, and another at the folded state[9], $Q = 1$. As before in the discussion of the small proteins these states will be separated by a kinetic barrier, hence by energetic and entropic contributions. If we neglect the entropy of the folded state to a first approximation, we have

$$F_{native} = E_N. \qquad (2.40)$$

At the folding temperature T_F, the free energies of the folded and unfolded state coincide

$$F_{native}(Q = 1, T_F) = F(Q_{min}, T_F) \qquad (2.41)$$

where the value of Q_{min} is close to that of the unfolded state. From this we can compute the *stability gap*

$$\delta E_s \equiv \overline{E}(Q_{min}) - E_N = S_0 T_F + \frac{\Delta E^2(Q_{min})}{2k_B T_F}. \qquad (2.42)$$

The folding temperature T_F can be related to the *glass transition temperature*, at which the entropy of the system vanishes, $S(E_0, Q) = 0$. It is given by

$$T_G(Q) = \sqrt{\frac{\Delta E^2(Q)}{2k_B S_0(Q)}}, \qquad (2.43)$$

with $S_0 = k_B \ln \Omega_0$ with Ω_0 given before.

The transition into the *glassy state* occurs precisely when there are too few states available, and the system remains frozen in one of those states. The ratio of the folding to the glass temperature is approximately given by

$$\frac{T_F}{T_G} \approx \frac{\delta E_s}{\sqrt{\Delta E^2}} \cdot \sqrt{\frac{2k_B}{S_0}}. \qquad (2.44)$$

In order for a protein to fold correctly and not be caught frozen in a wrong minimum, one needs to fulfill the condition

$$T_F > T_G. \qquad (2.45)$$

---

[9]The discussion of the small proteins indicates how to extend this idea to more available states.

Since S_0, ΔE^2 and E_N all depend linearly on N, the ratio Eq. (2.44) is not dependent on protein length, but very sensitive to interaction energies. This explains why the folding process remains so difficult to capture quantitatively.

Beyond this basic picture given here, there is still no general consensus on how to quantitatively model and predict protein folding; maybe there is no such theory ever to have, since such a great number of details intervene: proteins are individuals designed for very specific purposes, and this individuality may also be reflected in the way they fold. In the Additional Notes some interesting novel conceptual avenues to attack the folding problem are listed. The problem is likely to stay with us for some more time.

Here, we now move on to the problem of protein docking.

The *protein docking problem* as a specific bioinformatics problem dates back to the late seventies (J. JANIN and S. J. WODAK, 1978) and has since seen a continuous development towards more efficient algorithms. The term protein docking encompasses two closely related situations: *protein-protein docking*, as it occurs during the formation of protein complexes, and *protein-ligand docking*, which relates to the drug design problem mainly driving the field. Here, protein docking is an essential tool to help design specific and efficient small drug molecules.

In this section, we discuss the two main statistical physics aspects of the problem. In order to fit two molecules snugly together, they must certainly fit *geometrically*, but also their *interactions* must be compatible. While in fact both geometry and interaction are fundamentally coupled, the approximate approaches that exist to the docking problem today usually separate these questions in two independent procedures. *Geometric fit* is used as a first level of screening for docking candidates, and only then is the quality of the docking evaluated (*'scored'*) by a quantitative measure of the physical interaction of the docking partners.

Geometric fit. In order to determine the geometrical fit of two molecules, a quantitative method to determine surface complementarity is needed. We explain the algorithm introduced in (E. KATCHALSKI-KATZIR et al., 1992). It is based on the projection of the molecules in their atomic coordinates on a 3-dimensional grid built from N^3 grid points. The molecules are represented by discrete coordinates

$$a_{\mathbf{k}}^i = \left\{ \begin{array}{ccc} 1 & \epsilon & \Gamma \\ \varrho^i & \epsilon & \Omega \\ 0 & \epsilon & \Sigma \end{array} \right. \tag{2.46}$$

where $i = 1, 2$ represents the molecule of volume Ω and surface Γ, $\mathbf{k} = (l, m, n)$ are the lattice indices. A lattice point is considered inside the molecular volume

Ω if there is at least one atom within a distance r from it, where r is a parameter on the order of one *van der Waals atomic radius* - a measure for the short-range repulsion of the electron shells. The surface layer Γ then is a finite layer between the inside of the molecule and its exterior, Σ.

Matching of the surfaces is achieved by calculating correlation functions. The correlation between the a^i is given by

$$c_{\mathbf{q}} = \sum_{k=1}^{N} a_{\mathbf{k}}^1 \cdot a_{\mathbf{k+q}}^2 \qquad (2.47)$$

where $\mathbf{q}$ denotes the number of grid steps by which molecule a^2 is shifted with respect to molecule a^1. If the molecules have no contact, $c_{\mathbf{q}} = 0$. In order to penalize interpenetration of the molecules, ϱ^i will be assigned a large negative value for $i = 1$, and a small positive value for $i = 2$, leading to an overall negative contribution in $c_{\mathbf{q}}$. In this way, positive and negative correlations can clearly be related to the geometric positioning of the molecules.

Computationally, the determination of $c_{\mathbf{q}}$ would require the calculation of N^3 multiplications and additions for each of the N^3 relative shifts $\mathbf{q}$, i.e. N^6 steps. This can be reduced by taking advantage of the lattice representation. Calculating first the discrete Fourier transform of the $a_{\mathbf{q}}^i$, which can be done with the Fast Fourier Transform (FFT), and then going back by an inverse Fourier transform, the number of computational steps can be reduced to at most $\mathcal{O}(N^3 \ln N)$.

Docking energetics. The molecules typically have to find their partners in solution. As a consequence, the first quantity to determine is the *free energy of solvation* for each individual molecule. The Gibbs free energy for this process can be written as

$$\Delta G \equiv G^{solvent} - G^{vacuum} = \Delta G^{polar} + \Delta G^{nonpolar} . \qquad (2.48)$$

The nonpolar contribution consists of various terms, which are usually ascribed to cavity formation (i.e., to account for the cost of digging a hole into water to fit in the protein), conformational contributions, and van der Waals interactions. It is very difficult and debated how these quantities can be modelled; in any case, it is an approximate procedure. Here, we will be interested in electrostatic interactions only, which dominates the polar part in Eq. (2.48). For the solvation process shown in Figure 2.14, the relevant quantity is the *free energy of binding*. It is given by (R.M. JACKSON and M.J.E. STERNBERG (1995), A. HILDEBRANDT, (2005)),

$$\Delta G^{bind} = \Delta\Delta G^{solv,A} + \Delta\Delta G^{solv,B} + \Delta\Delta G^{solv,A-B} \qquad (2.49)$$

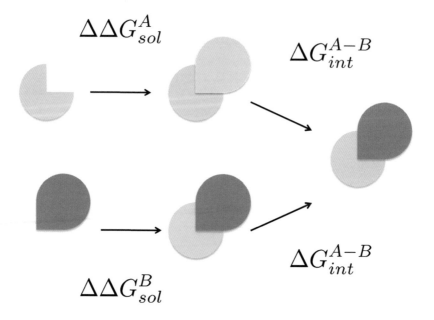

Figure 2.14 (**See color insert.**) Solvation of a protein complex in water; for explanation, see text.

The three contributions arise from: i) the change in solvation energy of molecule A upon binding; ii) the change in solvation energy of molecule B upon binding; iii) the interaction energy of A and B in the presence of solvent. The last contribution contains the energy stored in the bonds between both proteins, and in principle both polar and non-polar contributions.

This discussion obviously did not go beyond stating the basic principles. In Chapter 4 we will discuss how the electrostatic properties of biomolecules can be determined.

2.6 PULLING DNA

As a final topic for the application of the methods of statistical mechanics for biomolecules we introduce a model description for their elastic behaviour. Molecular elasticity is *entropic* in origin. This merits an explanation.

Let's begin with a simplistic version of a polymer chain model which is called the *freely jointed chain*. We imagine the polymer can be abstracted as a chain of N freely jointed links, as illustrated in Figure 2.15. The spatial configuration of the chain resembles the trajectory of a *random walk*.

Figure 2.15 The simplistic polymer model: a chain of linear segments which can freely move about their links.

If we now pull at the ends of our caricature polymer with a force **F**, we will straighten out the chain. What happens if we then let go? Imagining we have realized the chain as made from paperclips, it would simply stay straight, as there is no internal or external mechanism that would pull the segment back into the initial random configuration. Something is missing, this idea is wrong.

A biomolecule in solution is subjected to thermal energy, k_BT. The positional fluctuations of the surrounding molecules hit the chain and will thereby gently randomize the configuration. Straightening out the chain against these thermally fluctuating forces consequently reduces the entropy of the chain, i.e., the number of conformations it can adopt. Work has to be performed, and the relation between force and extension of a molecule in solution results thus from an *entropic elasticity*.

Based on this concept, various experimental techniques have been developed to exert forces on molecules in solution. The basic principle is that the molecule has to be fixed at one end, while the other can be manipulated by mechanical force. A typical setup is that a DNA molecule is chemically modified at its both extremities; it is then rigidly fixed at one end to a surface (a capillary), and to a superparamagnetic bead at the other end. The application of a magnetic field allows to pull on the bead and hence the DNA molecule. Another possible setup is by trapping a bead in an optical trap; such a setup is indicated schematically in Figure 2.16. Questions of biological or biophysical interest that have been addressed with this approach are listed in the additional notes at the end of this chapter; here we only want to understand how to describe the elastic properties.

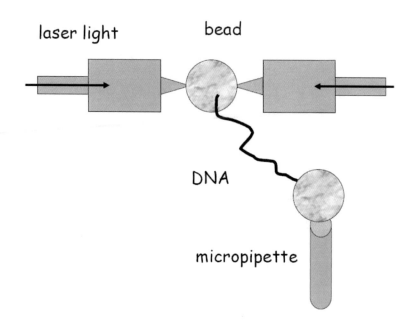

Figure 2.16 A setup for DNA stretching experiments using an optical trap.

The worm-like chain (WLC)-model. Following the suggested discrete representation of the polymer as a chain of elements, we first consider the polymer as a linear chain with element length b, such that the chain made of N segments has a total length of $L = bN$. The energy of the chain is given by the Hamiltonian

$$\beta H = -K \sum_i \widehat{\mathbf{t}}_i \cdot \widehat{\mathbf{t}}_{i+1} - \beta F b \sum_i \widehat{\mathbf{t}}_i \cdot \widehat{z} \tag{2.50}$$

where an applied force F is directed along the z-axis.

In Eq. (2.50), K/β is the *stiffness* of the chain, i.e., its resistance to bending. The unit vector $\widehat{\mathbf{t}}_i$ describes the orientation of the segment i. This model is sometimes called the *Kratky-Porod model*, which for $K = 0$ reduces to the *freely jointed chain* we had pictured before.

In the continuum limit $b \to 0$, where $K \equiv \ell_P/b$ introduces the *persistence length* ℓ_P, we can represent the model as the *Worm-like Chain model* with the energy expression

$$\beta H_{WLC} = \frac{\ell_P}{2} \int_0^L ds (\partial_s \widehat{\mathbf{t}})^2 - f \int_0^L ds\, \widehat{\mathbf{t}} \cdot \widehat{z} \tag{2.51}$$

where the pulling force per unit length is $F = k_B T f$. In this formulation, s is the curvilinear coordinate along the chain, and $\widehat{\mathbf{t}}$ has turned from the orientational vector of a chain element into the tangential vector along the continuous curve which makes out the chain.

The first term in the Hamiltonian is the *bending energy*. Its coefficient, the persistence length, is a length scale measure for the correlations of the tangent vectors along the chain, or, in other words, for the propagation of changes in the chain conformation along the chain.[10]

As stated above, the quantity we are interested in from an experimental point of view is the force-extension relation. For the WLC model, the exact computation of the partition function is equivalent to solving the problem of a quantum rotator subjected to a polarizing field. Here we give a simplified calculation valid for the case when the forces pulling at the molecule are sufficiently strong and limit the chain excursions transverse to the pulling direction (J. F. MARKO and E. SIGGIA, 1995).

Since $|\widehat{\mathbf{t}}| = 1$, we can take the transverse components $(t_x, t_y) \equiv \widehat{\mathbf{t}}_\perp$ as independent variables, and obtain for the z-component

$$t_z = 1 - \widehat{\mathbf{t}}_\perp^2/2 + O(\widehat{\mathbf{t}}_\perp^4). \tag{2.52}$$

We can then express the energy in a Gaussian approximation as

$$\beta H_{WLC} = \frac{1}{2} \int_0^L ds [\ell_p(\partial_s \widehat{\mathbf{t}}_\perp)^2 + f \widehat{\mathbf{t}}_\perp^2] - fL. \tag{2.53}$$

Introducing the Fourier modes

$$\widetilde{\mathbf{t}}_\perp(q) = \int ds\, e^{iqs} \widehat{\mathbf{t}}_\perp(s) \tag{2.54}$$

we obtain

$$\beta H_{WLC} = \frac{1}{2} \int \frac{dq}{2\pi} [\ell_p q^2 + f] \widetilde{\mathbf{t}}_\perp^2 - fL. \tag{2.55}$$

---

[10]There are several other commonly used lengths that characterize polymers; a prominent one is the *Kuhn length* with $\ell_K = 2\ell_p$, which is well defined provided $L \gg \ell_K$.

The average of the transverse components along the chain s follows from equipartition

$$\langle \widehat{\mathbf{t}}_\perp^2 \rangle = \int \frac{dq}{2\pi} \langle |\widetilde{\mathbf{t}}_\perp(q)|^2 \rangle = 2 \int \frac{dq}{2\pi} \left[\frac{1}{\ell_P q^2 + f} \right] = \frac{1}{\sqrt{f\ell_P}} \tag{2.56}$$

where the factor of two accounts for the two components of $\widehat{\mathbf{t}}_\perp$. The extension of the chain is then obtained as

$$\frac{z}{L} = \widehat{\mathbf{t}} \cdot \widehat{\mathbf{z}} = 1 - \frac{\langle \widehat{\mathbf{t}}_\perp^2 \rangle}{2} = 1 - \frac{\sqrt{f\ell_P}}{2} \tag{2.57}$$

which behaves as a square-root in f. The approximate formula

$$f = \frac{1}{\ell_P} \left[\frac{z}{L} + \frac{1}{4(1 - z/L)^2} - \frac{1}{4} \right] \tag{2.58}$$

describes the experimental behaviour rather well, see Figure 2.17.

Bending and twisting elasticity of DNA. J.F. MARKO and E.D. SIGGIA in 1994 proposed another continuum model for DNA elasticity which has enjoyed some recent interest also from the experimentalists; it is therefore briefly discussed at this point. The idea of this model is to look at the bending and twisting degrees of freedom of DNA and how they are coupled. Starting point is the schematics of Figure 2.18, in which a solid cylinder is shown on whose surface the sugar-phosphate backbone of the double-helix of DNA is inscribed, with orientations. Major and minor grooves are indicated by the letters **M** and **m**, respectively.

The orientation of the DNA is described by the set of vectors $(\mathbf{u}, \mathbf{v}, \mathbf{t})$ which forms a right-handed coordinate system at each point s along the curve, where s is arc length; see Figure 2.18. The space curve of DNA $\mathbf{r}(s)$ along the core of the cylinder hereby defines the tangential vector $\mathbf{t} = d\mathbf{r}/ds$ of unit length, $\mathbf{t}^2 = 1$. The vector $\mathbf{u}$ is chosen to lie at the middle of the intersections $\mathbf{R}$ and $\mathbf{S}$ that the two DNA strands make with a perpendicular plane across the cylinder. The third vector is then defined via the condition $\mathbf{v} = \mathbf{t} \times \mathbf{u}$.

With the definition of $\mathbf{e}^{(i)} = \mathbf{u}, \mathbf{v}, \mathbf{t}$ for $i = 1, 2, 3$, one can define a general deformation of the molecule by infinitesimal rotations $\Omega(s)$ of the coordinate axes according to the formula

$$\frac{d\mathbf{e}_i}{ds} = (\omega_0 \mathbf{e}^{(3)} + \Omega) \times \mathbf{e}^{(i)}, \tag{2.59}$$

where $\omega_0 = 2\pi/\ell$ where $\ell = 34$ Å is the helical pitch. If $\Omega = 0$, the configuration is undistorted as in Figure 2.18.

Figure 2.17 (**See color insert.**) Force-extension curve for a single-molecule DNA. (Reprinted from C. BUSTAMANTE et al., 2000, with permission from Elsevier.)

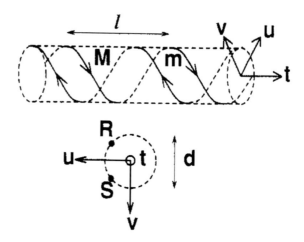

Figure 2.18 Geometry of the Marko-Siggia twist-bend model. [Reprinted from J.F. MARKO and E.D. SIGGIA (1994) with permission from the American Chemical Society.]

One can then write down a free energy expression as a Taylor expansion in small Ω and its derivatives. The result is

$$\frac{F}{k_BT} = \int_0^L ds \left[A_0 + \frac{1}{2}\sum_{ij} A_{ij}\Omega_i\Omega_j + \frac{1}{6}\sum_{i,j,k} A_{ijk}\Omega_i\Omega_j\Omega_k + ... \right]. \quad (2.60)$$

Truncating this expression at second order one finds on symmetry grounds that only six independent components survive (*Exercise!*) and the result then is

$$\frac{F}{k_BT} = \frac{1}{2}\int_0^L ds \left[\sum_{i=1,2,3} A_{ii}\Omega_i^2 + 2A_{23}\Omega_2\Omega_3 \right]. \quad (2.61)$$

This expression can be simplified by using the *Frenet-Serret equations*

$$\mathbf{t_s} = \kappa\mathbf{n}\,,\ \mathbf{n}_s = -\kappa\mathbf{t} + \tau\mathbf{b}\,,\ \mathbf{b}_s = -\tau\mathbf{n} \quad (2.62)$$

where the curvature $\kappa(s)$ and the torsion $\tau(s)$ have been introduced. Since the vector pairs $(\mathbf{u}, \mathbf{v})$ and $(\mathbf{n}, \mathbf{b})$ are coplanar, one can write

$$\mathbf{u} + i\mathbf{v} = e^{-i\phi(s)}(\mathbf{n} + i\mathbf{b})\,, \quad (2.63)$$

where the angle $\phi(s)$ rotates the slowly varying $(\mathbf{n}, \mathbf{b})$ into the rapidly rotating $(\mathbf{u}, \mathbf{v})$. The three variables κ, τ and ϕ can now be used to replace the three Ω_i. The final result then is given by

$$\frac{F}{k_B T} = \int_0^L ds \left[A\kappa^2 - B\kappa^2 \cos 2\phi + C(\phi_s - \overline{\omega})^2 + 2D\kappa(\phi_s - \overline{\omega}) \cos \phi \right] ,$$

(2.64)

with $A = (A_{11} + A_{22})/2$, $B = (A_{11} - A_{22})/2$, $C = A_{33}$ and $D = A_{23}$, and $\overline{\omega} = \Omega_0 - \tau$.

In this expression, we see that A is the usual bending rigidity as in the Kratky-Porod model, B is an asymmetric bending constant, C is the twist rigidity, and D is the twist-bend coupling. It is the latter that has been of recent interest to experimentalists, as the tweezer experiments are now becoming sufficiently sensitive to probe the presence of this coupling (S. K. NOMIDIS et al., 2017 and 2018; E. SKORUPPA et al., 2018).

Additional Notes

RNA structure. For RNA structure prediction, a large number of dedicated servers exists meanwhile; see the detailed catalogue on the web under

https://en.wikipedia.org/wiki/List_of_RNA_structure_prediction_software

H. ORLAND and his colleagues have further pursued the statistical physics approach to RNA structure and in particular looked at pseudoknots (M. PILLS-BURY at al., 2005), (M. BON et al., 2013), (G. VERNIZZI et al., 2016) and also true knots. The latter appear still to be absent in RNA, see (A.S. BURTON et al., 2016).

Protein folding and docking. A modern standard reference on the protein folding problem is the book by A. BEN-NAIM (2013). Otherwise there is an unconquerable amount of literature on protein folding. A. BEN-NAIM's second book on the topic which appeared in 2016 gives an account of this fact.

A recent review of protein docking methods is by N.S. PAGADALA et al. (2017).

Single-molecule pulling experiments. Experiments on single-molecule manipulations of nucleic acids have started in the early nineties of the last century; basic references on DNA elasticity are T. STRICK et al., (1996, 1998). Early reviews on the basic physics and the experimental techniques are C. BUSTAMANTE et al. (2000), and U. BOCKELMANN (2004).

The advantage of the single-molecule technique is that the mechanical force exerted on the molecule can be made to interfere with enzymatic reactions on the chains, allowing for a mechanical probing of biochemical processes. Examples from the catalogue of achievements are:

- Unwinding of promoter regions in transcription initiation: A. REVJAKIN et al. (2004).

- Pausing of RNA polymerase during transcription: R.J. DAVENPORT et al. (2000) and N.R. FORDE et al. (2002).

- Base-pair stepping of RNA polymerase: E.A. ABBONDANZIERI et al. (2005).

- Torque-induced structural transitions in DNA, of relevance for enzymatic reactions: Z. BRYANT et al. (2003).

- DNA unzipping: C. DANILOWICZ et al. (2003).

The field has continued to bring in new insights in ever more complex configurations and is now known under the name of *Single-Molecule Force Spectroscopy*. A recent review of the field of single-molecule techniques in biophysics is by H. MILLER et al. (2018).

Theoretical papers on DNA elasticity are the classic papers on the worm-like chain model by J.F. MARKO and E.D. SIGGIA (1995), and their earlier paper on the twist-bend coupling in 1994. Early extensions to include twist into the WLC model were published by C. BOUCHIAT and M. MÉZARD (1998), and J.D. MOROZ and P. NELSON (1997). This field has continued to develop in many directions, however, with the increasing use of molecular simulation approaches, into the direction of more detailed models on different level of atomistic detail - a topic this book does not address.

References

E.A. Abbondanzieri, W.J. Greenleaf, J.W. Shaevitz, R. Landick, S.M. Block
Direct observation of base-pair stepping by RNA polymerase
Nature **438**, 460-465 (2005)

B. Alberts, A. Johnson, J. Lewis, D. Morgan, M. Raff, K. Roberts, P. Walter
The Molecular Biology of the Cell
6th ed., Norton & Company (2014)

A. Ben-Naim
The protein folding problem and its solutions
World Scientific (2013)

A. Ben-Naim
Myths and Verities in Protein Folding Theories
World Scientific (2016)

U. Bockelmann
Single-molecule manipulation of nucleic acids
Curr. Op. Struct. Biol. **14**, 368-373 (2004)

M. Bon, C. Micheletti, H. Orland
McGenus: a Monte Carlo algorithm to predict RNA secondary structures with pseudoknots
Nucl. Acids. Res. **41**, 1895-1900 (2013)

C. Bouchiat, M. Mézard
Elasticity model of a supercoiled DNA molecule
Phys. Rev. Lett. **80**, 1556-1559 (1998)

Z. Bryant, M.D. Stone, J. Gore, S.B. Smith, N.R. Cozzarelli, C. Bustamante
Structural transitions and elasticity from torque measurements on DNA
Nature **424**, 338-341 (2003)

A.S. Burton, M. Di Stefano, N. Lehman, H. Orland, C. Micheletti
The elusive quest for RNA knots
RNA Biology **13**, 134-139 (2016)

C. Bustamante, S.B. Smith, J. Liphardt and D. Smith
Single-molecule studies of DNA mechanics
Curr. Op. Struct. Biology **10**, 279-285 (2000)

H.S. Chan and K.A. Dill
Protein folding in the landscape perspective: chevron plots and non-Arrhenius kinetics
Proteins **30**, 2-33 (1998)

C. Danilowicz, V.W. Coljee, C. Bouzigues, D.K. Lubensky, D.R. Nelson, M. Prentiss
DNA unzipped under a constant force exhibits multiple metastable intermediates
Proc. Natl. Acad. Sci. USA **100**, 1694-1699 (2003)

R.J. Davenport, G.J.L. Wuite, R. Landick, C. Bustamante
Single-molecule study of transcriptional pausing and arrest by E. coli RNA polymerase
Science **287**, 2497-2500 (2000)

B. Derrida
Random-Energy Model: limit of a family of disordered models
Phys. Rev. Lett. **45**, 79-82 (1980)

S.R. Eddy
How do RNA folding algorithms work?
Nat. Biotech. **22**, 1457-1458 (2004)

A. Fire, D. Albertson, S.W. Harrison, D.G. Moerman
Production of antisense RNA leads to effective and specific inhibition of gene expression in C. elegans muscle
Development **113**, 503-514 (1991)

N.R. Forde, D. Izhaky, G.R. Woodcock, G.J.L. Wuite, C. Bustamante
Using mechanical force to probe the mechanism of pausing and arrest during continuous elongation by Escherichia coli RNA polymerase
Proc. Natl. Acad. Sci. USA **99**, 11682-11687 (2002)

A. von Haeseler, D. Liebers
Molekulare Evolution
Fischer (2003)

A. Hildebrandt
Biomolecules in a Structured Solvent
Rhombos-Verlag (2005)

R.M. Jackson, M.J.E. Sternberg
A continuum model for protein-protein interactions: application to the docking problem
J. Mol. Biol. **250**, 258-275 (1995)

J. Janin and S.J. Wodak
Computer analysis of protein-protein interactions
J. Mol. Biol. **124**, 323-342 (1978)

E. Katchalski-Katzir, I. Shariv, M. Eisenstein, A.A. Friesem, C. Afaflo, I.A. Vakser
Molecular surface recognition: determination of geometric fit between proteins and their ligands by correlation techniques
Proc. Natl. Acad. Sci. USA **89**, 2195-2199 (1992)

C. Levinthal
How to fold graciously
in *Mössbauer Spectroscopy in Biological Systems*, J. C. M. Tsibris and P. Debrunner (eds.), 22-24 (1969)

J.F. Marko, E.D. Siggia
Bending and twisting elasticity of DNA
Macromolecules **27**, 981-988 (1994)

J.F. Marko, E.D. Siggia
Stretching DNA
Macromolecules **28**, 8759-8770 (1995)

M.L. Mehta
Random Matrices
2nd ed., Academic Press (1991)

H. Miller, Z.K. Zhou, J. Shepherd, A.J.M. Wollman, M.C. Leake
Single-molecule techniques in biophysics: a review of the progress in methods and applications, Rep. Prog. Phys. **81**, 024601 (2018)

J.D. Moroz, P. Nelson
Torsional directed walks, entropic elasticity, and DNA twist stiffness
Proc. Natl. Acad. Sci. USA **94**, 14418-14422 (1997)

S.K. Nomidis, F. Kriegel, W. Vanderlinden, J. Lipfert, E. Carlon
Twist-bend coupling and the torsional response of double-stranded DNA
Phys. Rev. Lett. **118**, 217801 (2017)

S.K. Nomidis, E. Skoruppa, E. Carlon, J. F. Marko
Twist-bend coupling and the statistical mechanics of DNA: perturbation theory and beyond
arXiv:1809.07050 (2018)

R. Nussinov, A.B. Jacobson
Fast algorithm for predicting the secondary structure of single-stranded RNA
Proc. Natl. Acad. Sci. USA **77**, 6309-6313 (1980)

J.N. Onuchic, Z. Luthey-Schulten, P.G. Wolynes
Theory of protein folding: The energy landscape perspective
Annu. Rev. Phys. Chem. **48**, 545-600 (1997)

H. Orland, A. Zee
RNA folding and large N Matrix theory
Nucl. Phys. B **620**, 456-476 (2002)

N.S. Pagadala, K. Syed, J. Tuszynski
Software for molecular docking: a review
Biophys Rev. **9**, 91-102 (2017)

M. Pillsbury, H. Orland, A. Zee
A steepest descent calculation or RNA pseudoknots
Phys. Rev. E **72**, 011911 (2005)

A. Revjakin, R.H. Ebright, T. R. Strick
Promoter unwinding and promoter clearance by RNA polymerase: Detection by single-molecule DNA nanomanipulation
Proc. Natl. Acad. Sci. USA **101**, 4776-4780 (2004)

E. Skoruppa, S.K. Nomidis, J.F. Marko, E. Carlon
Bend-induced twist waves and the structure of nucleosomal DNA
Phys. Rev. Lett. **121**, 088101 (2018)

T.R. Strick, J.-F. Allemand, D. Bensimon, A. Bensimon, V. Croquette
The elasticity of a single supercoiled DNA molecule
Science **271**, 1835-1837 (1996)

T.R. Strick, J.-F. Allemand, V. Croquette, D. Bensimon
Physical approaches to the study of DNA
J. Stat. Phys. **93**, 647-672 (1998)

G. Vernizzi, H. Orland, A. Zee
Prediction of RNA pseudoknots by Monte Carlo simulations
preprint q-bio/045014 (2004)

G. Vernizzi, H. Orland, A. Zee
Classification and predictions of RNA pseudoknots based on topological invariants
Phys. Rev. E **94**, 042410 (2016)

P.G. Wolynes
Landscapes, funnels, glasses and folding: from metaphor to software
Proc. Americ. Philosoph. Soc. **145**, 555-563 (2001)

M. Zuker
On finding all suboptimal foldings of an RNA molecule
Science **244**, 48-52 (1989)

CHAPTER 3

Phase Transitions in RNA and DNA

3.1 RNA PHASE BEHAVIOUR

In the previous sections we have seen that one can consider the RNA folding problem as one of combinatorics and energetics: the correct fold of an RNA molecule is the one which minimizes the free energy. All we have to do is put in the specific base pairing energies, starting from the Hartree equation (2.24).

However, even if we do this, a more fundamental problem arises. While for DNA Watson-Crick pairing of two complementary strands always gives a well-defined minimum free energy configuration, for RNA that is not clear. If the strands become longer and longer, the number of possible structures explodes. And many of these configurations can be at least almost degenerate: there will be several configurations with the same or nearly the same free energy. How accessible - and how relevant - is thus the minimum energy fold? Maybe we need to know also those structures that are not minimal in free energy, but close? Isn't it possible that the minimum configuration is well-separated from a nearby configuration by an energy barrier, and the molecular configuration, once trapped in the other minimum, will never change into the true minimum? Such a high barrier would exist, e.g., if the two energetically competing structures would be structurally far apart in configuration space. Thus, for RNA we come back to a problem which is very similar to that of protein folding and also here we have to discuss the free energy landscape of the RNA structures.

The free-energy landscape of RNA turns out to be strongly temperature-dependent. For high enough temperatures - but still below the denaturation temperature of a folded RNA, since we do not want to break the structure - one may be allowed to ignore sequence-dependence to a first approximation, as we did before.[1] We want to do this for the planar (non-pseudoknotted) structures and go back to the saddle-point approximation.

Under the assumption that sequence-dependence is negligible, the Hartree equation (2.24) becomes translationally invariant along the chain of length L and takes on the simplified form

$$G(L+1) = G(L) + q \sum_{k=1}^{L-1} G(k)G(L-k) \qquad (3.1)$$

with $q \equiv exp(-\beta \epsilon_0)$. By using the transform

$$\widehat{G}(z) = \sum_{L=1}^{\infty} G(L) z^{-L} \qquad (3.2)$$

the convoluted sum in Eq. (3.1) can be eliminated, which gives rise to an algebraic equation for $\widehat{G}(z)$,

$$z\widehat{G}(z) - 1 = \widehat{G}(z) + q\widehat{G}^2(z). \qquad (3.3)$$

This equation has the solution

$$\widehat{G}(z) = \frac{z - 1 - \sqrt{(z-1)^2 - 4q}}{2q}. \qquad (3.4)$$

In order to transform back to $G(L)$, again a saddle-point approximation can be used. We then find the expression

$$G_0(L) \approx g_0(q) L^{-3/2} exp(-Lf_0(q)) \qquad (3.5)$$

with $f_0(q) = -\ln(1 + 2\sqrt{q})$.

The result (3.5) shows the coexistence of an exponentially large number of RNA secondary structures with *equal* free energy. This phase is dominated by the configurational entropy of the molecules; it has been termed the '*molten phase*' (R. BUNDSCHUH and T. HWA, 2002).

How physically realistic is this phase? In order to check this, one has to vary the binding energies. In doing so we introduce what can be called *sequence disorder*, since we start from a molecule with identical bindings. We choose binding energies ε_{ij} with

$$\varepsilon_{i,j} = \begin{cases} -u_m & (i,j) \ , \ WC \\ u_{mm} & , \quad else \end{cases} \qquad (3.6)$$

---

[1]But then based on a different argument since wanted to look at the different possible topologies.

with $u_m, u_{mm} > 0$, and WC stands for a Watson-Crick base. The $\varepsilon_{i,j}$ are taken as independent Gaussian distributed random variables with mean $\bar{\varepsilon}$ and variance

$$\overline{(\varepsilon_{i,j} - \bar{\varepsilon})(\varepsilon_{k,l} - \bar{\varepsilon})} = D\delta_{i,k}\delta_{j,l}. \tag{3.7}$$

For this choice it turns out that the molten phase is indeed thermodynamically stable for weak sequence disorder (R. BUNDSCHUH and T. HWA, 2002), but if we allow for arbitrarily strong sequence disorder the molten phase becomes unstable with respect to another phase: the *glass phase*. This phase is dominated by one or few structures of lower free energies.

How can one see that the molten phase must be unstable with respect to the glass phase? We follow an argument developed by R. BUNDSCHUH and T. HWA, 2002.

The argument is based on the introduction of a so-called *division free energy* in the molten phase, given by

$$\Delta F \equiv -k_B T \ln \left[\frac{G_0^2(L/2)}{G_0(L)} \right] \approx \frac{3}{2} k_B T \ln L. \tag{3.8}$$

It is the free energy cost of cutting a chain of length L into two non-interacting halves. This quantity characterizes the loss of configurational entropy a secondary structure of the molten phase will undergo by cutting.

Let us now consider the division free energy for an arbitrary base sequence. For each sequence, we select a segment $\ell \ll L$ of Watson-Crick-paired bases r_i such that the sequence $r_i...r_{i+\ell-1}$ is in the first half of the molecule, while $r_{j-\ell+1}...r_j$ is in the second half. For a random sequence of length L one can show rigorously that $\ell = \ln L / \ln 2$. We further observe that

$$\Delta F \equiv F_{div} - F_{free} \geq F_{div} - F_{paired} \tag{3.9}$$

where F_{paired} is the free energy of the ensemble of structures in which the complementary segments are paired and unpaired substrands remain,

$$F_{paired} = -\ell u_m + (N - 2\ell)f_0 + \frac{3}{2} k_B T [\ln L_1 + \ln L_2] \tag{3.10}$$

where L_1, L_2 are the lengths of the two remaining substrands. Taking as F_{div} the value of the molten phase,

$$F_{div} = f_0 L + 2(3/2)k_B T \ln L \tag{3.11}$$

(ignoring L-independent terms), and assuming that L_1, L_2 scale linearly with L, one is left with the estimate

$$\frac{3}{2} k_B T \geq \frac{u_m + 2f_0}{\ln 2}.$$

(3.12)

To close the argument we have to discuss the *low temperature limit* of Eq. (3.12). For $T \to 0$, $f_0 \to -\bar{n}u_m$, where $\bar{n}$ is the average number of base pairs per monomer in the minimal free energy structure; if all bases are paired, we have $\bar{n} = 1/2$. Since there will in general be a finite fraction of non-paired bases, we expect $\bar{n} < 1/2$. Thus, by increasing the value of u_m while decreasing T, the inequality (3.12) must be violated at a critical value u_m^*, necessitating the existence of a phase of secondary structures at a lower temperature. This then is the *'glass phase'*.

For real RNA sequences, apart from the glass phase, we also have to consider the *'native phase'*, which corresponds to the minimum free energy solution. One can say the structural phases of RNA are characterised by the average sizes of the molecules based on the scaling behaviour of the average diameter $\langle h \rangle$. Supposing a relation

$$\langle h \rangle \sim L^m$$

(3.13)

between diameter $\langle h \rangle$ and sequence length L, introducing an exponent m, one finds that the exponent varies between $0.5 < m < 1$ in the glass, molten and native structures. Thus, $\langle h \rangle$ displays only a rather weak dependence on L.

3.2   THE DNA MELTING TRANSITION

The most fundamental feature of DNA is its ability to hybridize its two complementary strands. This is the prime example of the physico-chemical recognition processes which are at the heart of essentially all biological mechanisms on a molecular level.

In some sense, the base-pairing mechanism of DNA is also the simplest of these recognition processes. It can be easily studied experimentally *in vitro*: one only needs to heat a solution containing double-stranded DNA molecules. The *thermal denaturation* or *melting* of DNA is the dissociation of the double-stranded DNA into its two separate single stands - hence just the inverse of the recognition mechanism. This process is reversible: by cooling the sample, the single strands rehybridize to the double strand. As first noticed by R. THOMAS in the early 1950's, this denaturation/hybridization process is indeed a thermodynamic phase transition in the limit of long chains. Many features of this phase transition remain correct in a more direct biological context, namely when the double strand of DNA is opened by the application of a direct force, e.g., by an enzyme.

We will now start to describe the thermal denaturation process, and we begin with short chains. This case is simpler than that of longer chains and also has important applications in biotechnology, e.g., for DNA microarrays to which we will turn later on as well.

Denaturing short chains. The opening of double-stranded DNA molecules has two aspects. Ignoring first any sequence-dependent effects, it can be understood as a simple dissociation process: as one double strand opens, it gives rise to two single strands. One can easily imagine that when the chains become longer and longer, the sequence-dependence will become more and more important, and dissociation will not simply be one double strand versus two single strands anymore; many intermediate states will become possible in which the double-stranded molecule is only partially dissociated.

As we have said, in the case of short chains (typically up to, say, 50 bp) we want to model the denaturation process as a dissociation equilibrium

$$C_2 \leftrightarrow 2C_1 \tag{3.14}$$

where C_1 is the concentration of single strands, and C_2 the concentration of double strands. This reaction is governed by the equilibrium constant $K_D = C_2/C_1^2$. The total concentration of DNA is given by $C_T = C_1 + 2C_2$, since each double strand contains two single strands. The quantities of interest are the fraction of double strands,

$$\theta_D \equiv \frac{2C_2}{C_T}, \tag{3.15}$$

or, likewise, the fraction of single strands, $\theta = 1 - \theta_D$. Using the definition of K_D, a quadratic equation for C_1 can be obtained which allows to express θ_D or θ in terms of K_D and C_T. The fraction of single strands is then found to fulfill the equation

$$\theta = \frac{-1 + \sqrt{1 + 2\gamma}}{\gamma}, \tag{3.16}$$

where $\gamma \equiv 4C_T/K_D$. This calculation assumes differing strands; if the two strands are self-complementary, $4C_T$ has to be replaced by C_T. *Why? Exercise.*

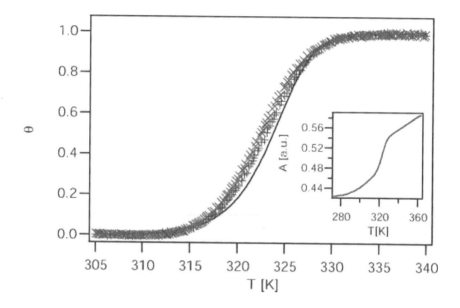

Figure 3.1 (**See color insert.**) Melting curve of a 16-bp oligomer; comparison of theory and experiment. [Reprinted with permission from J. BAYER et al. Copyright (2005) American Chemical Society.]

What remains to be calculated is the equilibrium constant K_D; in particular, its temperature dependence will turn out to be important. One has

$$K_D = \exp[-(\Delta H - T\Delta S)/RT] \qquad (3.17)$$

where the transition enthalpy given by

$$\Delta H = \Delta H_{H-bond} + \Delta H_{nn} \qquad (3.18)$$

contains both contributions from the hydrogen bonds between the paired bases on the two strands, and from the *stacking* of the pairs on top of each other, see the previous section. Typically, for the stacking contribution it is assumed that it can be considered as a purely nearest-neighbor effect, but there are situations (e.g., in the presence of mismatches) when this is not sufficient.[2]

---

[2]The same philosophy applies to RNA-RNA and RNA-DNA duplexes, but the values of the stacking interactions differ in all these cases.

The transition entropy contribution in Eq. (3.17) is typically approximated by $\Delta S = \Delta S_{bp}[N_{AT} + N_{GC}]$ where ΔS_{bp} is a temperature-independent value, and N_{AT} and N_{GC} are the number of AT- and GC-base pairs, respectively. The precise values of the melting parameters have been determined experimentally, see, e.g., R. OWCZARZY et al., 1998, and J. SANTA-LUCIA, 1998.

Figure 3.1 illustrates the comparison between theory and experiment for the sequence 5'-TAG TTG TGA TGT ACA T-3'. (J. BAYER et al., 2005). The graph illustrates generic features observed in oligomer melting curves. The DNA double-strand undergoes a melting transition at a temperature T_M which is at $\theta = 0.5$; here, the slope of the curve is maximal (the derivative $d\theta/dT$ has a maximum). The melting temperature T_M depends on base composition and strand length. Further, the shape of the melting curve turns out to be universal; it generically has a sigmoidal form. The effect of a change in base sequence thus leads to a shift in melting temperature while the overall curve is unchanged. If the base sequence is extended by adding more bases, the melting temperature will shift to higher temperatures - simply because more bases have to be broken. At the same time, however, the sigmoidal curve will steepen up. This is a consequence of a *cooperative effect*: as long as the denaturation process occurs in a two-state fashion, essentially all bases have to open up in concert.

Melting long chains: the Poland-Scheraga model. We now turn to the discussion of melting of long DNA chains which do not open in the simple two-state fashion. We can easily imagine that if we make the chains longer and longer, 50 bp to, say, 500 bp, the cooperative effect of all bases opening in concert will not be operative any more all along the chain. We may then expect that a DNA molecule undergoing the denaturation process may look like the schematic illustration in Figure 3.2. The molecule consists of a sequence of helices and open segments, so-called *denaturation loops* or *bubbles*. The configuration shown in addition has open ends, but closed ends are of course possible, too.

A model to describe the statistics of these configurations was first proposed by POLAND and SCHERAGA (1970); we here follow the discussion by Y. KAFRI et al. (2002).

If we ignore the configurational entropy of a bound segment embedded in the ambient space, its statistical weight will be given by

$$w^\ell = exp(-\ell E_0/k_B T) \tag{3.19}$$

where ℓ is the length of the segment, and E_0 is a binding energy in which, for the moment, no distinction is made between the chemical nature of the bases.

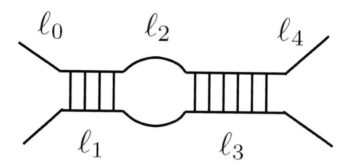

Figure 3.2 A DNA chain configuration near the melting transition: it consists of an open end of length ℓ_0, a bound segment of length ℓ_1, a loop of length ℓ_2, a bound segment of length ℓ_3, and an open end of length ℓ_4.

By contrast, a *denatured loop* has no energy associated with it; its statistical weight is consequently determined only by its degeneracy - we have to count the number of its configurations in order to estimate its contribution to the entropy of the configuration.

Assuming that the DNA loop is fully flexible, in the simplest modelling it can be considered as a random walk which returns to its origin after a path of length 2ℓ. From this modelling idea one then knows that the statistical weight for the loop of length ℓ has an algebraic form,

$$\Omega(2\ell) = \sigma \frac{s^\ell}{\ell^c} \qquad (3.20)$$

where s is some constant; the amplitude prefactor σ will be simply put to $\sigma = 1$ for the moment; but we will come back to it later.

The exponent c is determined by the statistical properties of the loop configurations. Finally, the configuration of the chain ends must be characterized; they consist of two denatured strands of length ℓ with a weight $\Lambda(2\ell)$ of a similar form as in Eq. (3.20), but with a different exponent $\bar{c}$. The values c and $\bar{c}$ can assume will be given later.

Given these weights, we can now proceed to calculate the total weight of any given configuration. Supposing the example shown in Figure 3.2, its statistical weight is obtained by

$$\Lambda(2\ell_0)w^{\ell_1}\Omega(2\ell_2)w^{\ell_3}\Lambda(2\ell_4). \tag{3.21}$$

The melting phase transition. This section is somewhat more formal since we now want to determine the nature of the transition between a bound and a denatured phase of a double-stranded DNA molecule.

In order to determine the thermodynamic properties of this model, it is practical for the calculation to assume a grand canonical ensemble in which the total chain length L is allowed to fluctuate.[3] The reason why this is a good choice is that in this ensemble the total partition function can be conveniently expressed as a geometric series,

$$\mathcal{Z} = \sum_{L=0}^{\infty} Z(L)z^L = \frac{V_0(z)Q(z)}{1 - U(z)V(z)}, \tag{3.22}$$

where $Z(L)$ is the canonical partition function of a chain of given length L, and z is the *fugacity*.[4] The functions $U(z)$, $V(z)$ and $Q(z)$ are defined as

$$U(z) \equiv \sum_{\ell=1}^{\infty} \Omega(2\ell)z^\ell = \sum_{\ell=1}^{\infty} \frac{(sz)^\ell}{\ell^c} = \Phi_c(sz), \tag{3.23}$$

$$V(z) \equiv \sum_{\ell=1}^{\infty} w^\ell z^\ell = \frac{wz}{1 - wz}, \tag{3.24}$$

$$Q(z) \equiv 1 + \sum_{\ell=1}^{\infty} \Lambda(2\ell)z^\ell = 1 + \Phi_{\bar{c}}(sz) \tag{3.25}$$

and

$$V_0(z) = 1 + V(z). \tag{3.26}$$

$\Phi_c(sz)$ is the *polylog function* which converges for $|z| < 1$; if in addition $\mathrm{Re}(c) > 0$, the function has an integral representation

$$\Phi_c(z) = \frac{1}{\Gamma(c)} \int_0^\infty dt\, t^{c-1} \frac{ze^{-t}}{1 - ze^{-t}}, \tag{3.27}$$

---

[3]Remember that we argued in Chapter 1 that in the thermodynamic limit the ensembles are equivalent.

[4]The fugacity is the intensive variable conjugate to the chain length, as much as the chemical potential is conjugate to the particle number.

where $\Gamma(c)$ is Euler's gamma function. The integral representation allows to see that $\Phi_c(z)$ has a divergence of the form $|z - 1|^{c-1}$ for $z \to 1$, if $c \leq 1$. If $c > 1$ and $1 - z = \varepsilon \ll 1$, one has $\Phi_c(1) - \Phi_c(1-\varepsilon) \sim \varepsilon^\zeta$ where $\zeta = \min(1, c-1)$.

In order to fix the average chain length L one has to choose the fugacity such that

$$L = \frac{\partial \ln \mathcal{Z}}{\partial \ln z}. \tag{3.28}$$

The thermodynamic limit $L \to \infty$ is obtained by allowing z to approach the lowest fugacity value z^* for which the partition function $\mathcal{Z}$ diverges. This divergence can have two sources: either the numerator grows without bounds, or the denominator vanishes. In fact, both cases occur. At low temperatures, the denominator vanishes, which is at

$$U(z^*)V(z^*) = 1. \tag{3.29}$$

Making use of expression Eq. (3.24) for $V(z)$, this result can be expressed as

$$U(z^*) = \Phi_c(sz^*) = \frac{1}{wz^*} - 1 = \frac{1}{V(z^*)}. \tag{3.30}$$

The solutions of this equation depend on the singularities of $\Phi_c(z)$, which depend itself on the value of c.

We take as the order parameter for the denaturation transition in the long chains the fraction of bound monomers[5], θ_b. Its temperature dependence in the thermodynamic limit can be calculated from the behaviour of $z^*(w)$, since the average number of bound pairs in a chain is given by

$$\langle m \rangle = \frac{\partial \ln \mathcal{Z}}{\partial w} \tag{3.31}$$

so that

$$\theta_b = \lim_{L \to \infty} \frac{\langle m \rangle}{L} = \frac{\partial \ln z^*}{\partial \ln w}. \tag{3.32}$$

In the case of the short chains, we defined the transition as the point where the temperature-derivative of the fraction of bound (or, unbound) base pairs would show a maximum. We also noted that for somewhat longer chains, this curve would steepen, and we interpreted this as a cooperative effect. Indeed, it is a signature of a collective phenomenon, which in the limit of infinite system becomes *sharp*, provided all bases were to open up collectively.

---

[5] For the case of short chains, $\theta_b = \theta_D$.

If we now look at the *order parameters* $\langle m \rangle$ and θ_b we defined for the state of the DNA double-strand, we are asked to check what the possible limiting behaviour of these quantities in the limit $L \to \infty$ is. This whole discussion resembles the (simpler) discussion of the phase transition in the 1-dimensional Ising model in Chapter 1. The difference between the two is that, apart from the energy of the configurations, we have to explicitly account for the *loop entropy*. And that, as we will see, makes everything different since the entropic weights of the loops are algebraic in nature, hence of a long-range nature and not rapidly decaying as for an exponential contribution.

As a consequence of this observation, the critical quantity in this discussion is the exponent c. Depending on its value we can find three different scenarios:

- $c \leq 1$: **no phase transition.** The function $U(z)$ is finite for all $z < 1/s$, and it diverges at $z = 1/s$. The function $1/V(z)$ is always finite and intersects $U(z)$ for $z < 1/s$. There is no singular behaviour.

- $1 < c \leq 2$: **continuous phase transition.** In this case, $U(z)$ is finite at $z = 1/s$ since $c > 1$. For $z > 1/s$, it is infinite. The singular point $z_M = 1/s$ is thus the phase transition point, separating a bound and a denatured regime. At the transition, the derivative of $U(z)$ diverges.

- $c > 2$: **first-order phase transition.** In this case, $U(z)$ and its derivative are finite at $z = z_M$. Again, there is a transition at this point; above the transition, θ_b vanishes in the thermodynamic limit. The transition is thus first-order.

All that remains to do now is to fix the value of c, which so far has been left unspecified. Since we have modeled the loops as random walks, we have to quantify the relationship between random walks and the polymeric nature of DNA.

Indeed, if we were to consider the DNA loop simply as a random walk of a given length which returns to the origin, we would find a value of $c = d/2$ in d space dimensions. Taking this exponent, and looking into the list of possible behaviours, there would be no denaturation transition in $d \leq 2$; for $2 < d \leq 4$ the transition is continuous, and for $d > 4$, the transition will be first-order.

This argument can be somewhat refined when considering the loop as a *self-avoiding walk*, since for a purely random walk configurations can occur in which the walk crosses itself. The assumption of self-avoidance seems more realistic since a real polymer can of course not intersect itself. One has

$$c = d\nu \tag{3.33}$$

where ν is the exponent associated with the *radius of gyration*[6] R_G of a self-avoiding random walk. For a walk of length L one has $R_G \sim L^\nu$ with $\nu = 3/4$ in $d = 2$ and $\nu \approx 0.588$ in $d = 3$. For the loop exponent c, this yields $c = 1.5$ in $d = 2$, and $c = 1.766$ in $d = 3$. The inclusion of self-avoidance thus leads to a slight smoothing of the transition.

All these reasonings are based on the assumption that the loop can be considered an isolated object which does not interact with the rest of the chain. In fact, the whole build-up of the weights assumes that one can consider the 'bits and pieces', i.e., the helices and loops, as essentially independent from each other. Recently, however, arguments have been put forward to account for the self-avoidance between the loops and the rest of the chain, albeit in an approximate way. These so-called *excluded-volume effects* between a loop and the chain have originally been derived using the theory of polymer networks by B. DUPLANTIER, 1986. The discussion of his theory goes beyond the scope of this book since it relies on renormalization group arguments; here, it must suffice us to simply state that the excluded volume interactions modify the self-avoidance exponent relation Eq. (3.33) into the expression

$$c = d\nu - 2\nu_3 \tag{3.34}$$

where ν_3 takes into account the contribution from the two 3-vertices at the extremities of a loop, which connect the loop to the chain. In $d = 3$, one finds a value of $\nu_3 \approx -0.175$. Thus for c results

$$c \approx 2.115 \tag{3.35}$$

i.e., a value which is slightly *larger* than two. This result is in accord with independent numerical work based on Monte-Carlo calculations for lattice models of DNA (E. CARLON et al., 2002).

A side remark on the boundary effects. Just in order to complete the discussion, it turns out that the result on the denaturation transition does not depend on the value of $\bar{c}$. This exponent enters the discussion, e.g., if one is interested in the average length of the end segment near the transition, which is given by

$$\xi = z \frac{\partial \ln Q}{\partial z} \bigg|_{z=z^*} \tag{3.36}$$

with the value of z^* determined before. The three different behaviours found for the denaturation transition can be found back for the average end segment

---

[6]The radius of gyration is the average distance of a monomer from the centre of mass of the polymer.

length; for more details of this behaviour, see Y. KAFRI et al., 2002. Here we simply note that the value of $\bar{c}$ can be obtained by summing up the scaling dimensions of a linear chain and a fork, leading to $\bar{c} = -(\nu_1 + \nu_3) \approx 0.092$, i.e., a very small value. Near the melting transition the end segment diverges according to

$$\xi \sim \frac{1}{|T - T_M|}, \tag{3.37}$$

where T_M is the melting temperature.

The melting transition with sequence dependence. The main result of the previous section was that the denaturation transition may in fact be a first-order rather than a continuous transition if excluded-volume effects between the bubbles and the chain do matter.

This finding has revived a dormant, but long-ongoing controversy on the nature of the transition. We will now turn to the computation of the melting curves of specific DNA sequences and want to see whether the computational results can faithfully represent experimental data. We thus first have to build in sequence-dependence into the theory which we had ignored so far.

In order to systematically build in sequence effects we will follow a version of the *Poland-Scheraga model* which is also useful for further generalizations (like the thermal stability of hybrid DNA, i.e., chains which are not fully complementary and/or of different length). T. GAREL and H. ORLAND (2004) have recently cast the Poland-Scheraga model into a recursive formulation based on the partition function.[7] Previous work by D. POLAND (1974) had relied on a recursion formulation for configuration probabilities. Based on these, the simulation program MELTSIM had been developed by R. BLAKE et al. (1999), a variant of which had been used to obtain the simulation results described here.

In the original Poland-Scheraga model one only considers two complementary strands of equal length N. Let us call $Z_f(\alpha)$ the *forward partition function* of the two strands starting at base 1, and ending at base α (which is assumed paired). The interaction of the bases is built as that between base pair stacks at positions α and $\alpha + 1$ along the chain, with associated stacking energies[8]

$$\epsilon_\alpha \equiv \epsilon_{\alpha,\alpha+1;\alpha,\alpha+1}. \tag{3.38}$$

---

[7]We have seen a basic version of this procedure in the section on the Ising model in Chapter 1, and will encounter another application in the discussion of RNA secondary structure in the following chapter.

[8]The values for these energy parameters are defined such as to include the hydrogen bond contributions.

Figure 3.3 Graphical representation of the Poland recursion for the partition function.

In order to find the recursion relation obeyed by $Z_f(\alpha+1)$ one has to consider the three possibilities to pair two bases at position $\alpha + 1$:

- the last pair $(\alpha, \alpha + 1)$ is stacked;

- there is a loop starting at any α' with $1 \leq \alpha' \leq \alpha - 1$ which ends at $\alpha + 1$;

- there is no loop.

This is illustrated in Figure 3.3. Formally, this recursion relation is expressed as

$$Z_f(\alpha + 1) = e^{-\beta \epsilon_\alpha} Z_f(\alpha) + \sigma \sum_{\alpha'=1}^{\alpha-1} Z_f(\alpha') \mathcal{N}(2(\alpha + 1 - \alpha')) + \Sigma \mathcal{M}(\alpha) \quad (3.39)$$

where $\beta = 1/kT$. There are two *cooperativity parameters* in this expression: σ and Σ; these parameters quantify the probabilities for loop and fork formation, respectively, and are assumed to be sequence-independent.

The factor $\mathcal{N}$ counts the number of conformations of a chain beginning at base α' and ending at $\alpha + 1$. Asymptotically, one has

$$\mathcal{N}(2(\alpha + 1 - \alpha')) = \mu^{\alpha - \alpha'} f(\alpha - \alpha') \quad (3.40)$$

where $k_B \ln \mu$ is the entropy per base pair (irrespective of its nature), and $f(\ell)$ is the return probability of a loop of length 2ℓ to the origin. Finally, in Eq. (3.39), $\mathcal{M}(\alpha) = \mu^\alpha g(\alpha)$, which counts the number of conformations of a pair of unbound chains starting at base 1 and paired at base $\alpha + 1$. The function g has a power-law behaviour, however, as we have seen before, with an exponent close to 0. We therefore simply put this factor to 1.

In a similar fashion as for the forward partition function Z_f one can compute the backward partition function Z_b, starting at base N and ending at a paired base α. Again there are three options to pair a base at position α, and thus one finds

$$Z_b(\alpha + 1) = e^{-\beta\epsilon_\alpha} Z_b(\alpha + 1) + \sigma \sum_{\alpha'=\alpha+2}^{N} Z_b(\alpha')\mathcal{N}(2(\alpha' - \alpha)) + \Sigma\mathcal{M}(N - \alpha).$$

(3.41)

From the expressions Eq. (3.39,3.41) one obtains the probability for the binding of a base pair α as

$$p_\alpha = \frac{Z_f(\alpha)Z_b(\alpha)}{Z},$$

(3.42)

where Z is the partition function of the two strands, given by

$$Z = Z_f(N) + \Sigma \left(\mu Z_f(N - 1) + \mu^2 Z_f(N - 2) + ... + \mu^{N-1}Z_f(1)\right)$$

(3.43)

or, expressed in terms of the backward partition function,

$$Z = Z_b(1) + \Sigma \left(\mu Z_b(2) + ...\mu^{N-1}Z_b(N)\right).$$

(3.44)

For the implementation of these recursion relations one has to take into account that the algorithm is $\mathcal{O}(N^2)$ since one has to compute $\mathcal{O}(\alpha)$ terms for each value α. In order to reduce the computational complexity, M. FIXMAN and J. J. FREIRE, 1977 have developed an approximation in which the power-law loop-entropy factor is replaced by a sum of exponentials

$$f(\ell) = \frac{1}{\ell^c} \approx \sum_{i=1}^{I} a_i e^{-b_i \ell}$$

(3.45)

where the I parameters (a_i, b_i) obey a set of non-linear equations and determine the degree of accuracy. With this step, computational complexity is reduced to $\mathcal{O}(N \cdot I)$. In the case of sequences of unequal length, the complexity of the algorithm can be reduced with this method from $O(N_1^2 N_2^2)$ to $O(N_1 N_2)$.

We can turn to the application of the PS-model, and we will see that the story that results will be, to some extent, a story of the cooperativity parameter σ.

3.3 THE MELTING PROFILES OF GENOMIC DNA AND CDNA

Melting genomic DNA. We now want to take a look at the comparison of the theory we have described with experimental data. In particular we ask what value of c, the one without excluded-volume or with inclusion of excluded-volume effects does fit experiments best? Could we even find out this way whether the transition is first- or second order?

In order to perform such a quantitative comparison, the theory has finally to be complemented by experimental parameters. The first set of parameters are the nearest and paired neighbor energies (R. D. BLAKE et al., 1999), distinguishing the different paired bases and their stacks; there are ten independent values to be specified. Further, there is the amplitude factor, the *cooperativity parameter* σ which we had put to a value of one before. On a technical level, this parameter determines the relative magnitude of energetic and entropic effects. Its name, however, indicates its physical meaning: it determines how many bases interact cooperatively, i.e., open up together to disrupt the helix and form a loop.

(a)

(b)

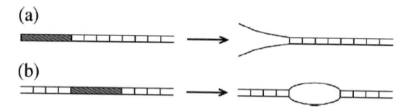

Figure 3.4 (**See color insert.**) Schematic diagram of the linearization of an AT-insertion in a GC-rich bacterial plasmid. The bar represents the inserted DNA segment which upon cutting is either placed at an extremity of the molecule (a) or at its center (b). [Reprinted with permission from R. BLOSSEY and E. CARLON. Copyright (2003). American Physical Society.]

Figure 3.5 Fits of the melting curve with different values of c and σ. The curve with $c = 2.15$ and $\sigma = 1.26 \times 10^{-5}$ does not fit the experimental data; by contrast, two choices for c and σ lie exactly on top of each other. [Reprinted with permission from R. BLOSSEY and E. CARLON. Copyright (2003). American Physical Society.]

A good system to test the issue of the order of the denaturation transition has been introduced by R. D. BLAKE and S. G. DELCOURT, 1998. They inserted artificial AT-rich sequences of varying length into GC-rich bacterial (plasmid) DNA, with sequence lengths between about 200 and 700 bp. After the insertion, the circular plasmid DNA was linearized in two ways: either the insertion was left at the extremity, or left imbedded in the GC-rich chain, see Figure 3.4. Both configurations differ in their melting temperatures: the sequence which melts from the extremity has a lower T_M, i.e., $T_M^{loop} > T_M^{end}$.

Figure 3.5 shows a first example of a calculated differential melting curve for the case of an embedded loop, for the longest inserted sequence with 747 bp, and for three sets of values (σ, c). It is found that the curves with the sets $(\sigma = 1.26 \times 10^{-5}, c = 1.75)$ and $(\sigma = 1.75 \times 10^{-4}, c = 2.15)$ fall on top of each other; the curve with the value $\sigma = 1.26 \times 10^{-4}$ and $c = 2.15$ has a higher T_M. Note that the two curves which fall on top of each other in this graph are in accord with experiment (data not shown). This result shows that a change in the value of c can apparently be compensated for by choosing a smaller value of σ.

Figure 3.6 Length dependence of the melting temperature shift. [Reprinted with permission from R. BLOSSEY and E. CARLON. Copyright (2003). American Physical Society.]

Figure 3.6 shows the theoretical values for the melting temperature difference $\Delta T_M \equiv T_M^{loop} - T_M^{end}$ the two sets (σ, c), in comparison with the experimental data, as a function of inverse insert length, $1/N$. Both theoretical curves deviate for shorter chain lengths, for which the theory is less reliable (see below). Given that the experimental resolution of melting temperatures is on the order of 0.1°C, there is obviously no direct way to decide the issue of the nature of the transition based on a comparison of the Poland-Scheraga model with this kind of experiment.

If one considers longer sequences of genomic DNA, such as, e.g., the human cancer-related gene eIF-4G with a sequence length of about 2900 bp, a structure-rich denaturation profile appears with many interior openings, hence loops, along the sequence. This is illustrated in Figure 3.7, where the differential melting signal is shown together with the denaturation probability $1 - A(i)$, which is the probability that the i-th base pair is in a closed state. The figure shows this probability at six different temperatures, labelled with Greek letters $\alpha, .., \phi$. Again, in this figure, the melting curves are compared for different values of σ and c, and as before for the plasmid inserts, again the

Figure 3.7 Melting curve of the cancer-related human gene eIF-4G with about 2900 bp (a); opening regions along the chains at different temperatures (b). [Reprinted with permission from R. BLOSSEY and E. CARLON. Copyright (2003). American Physical Society.]

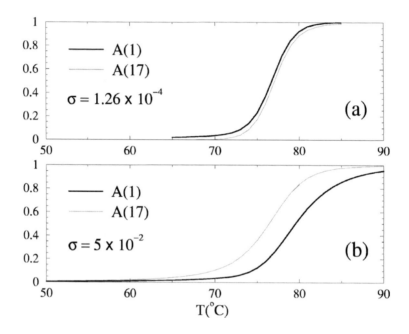

Figure 3.8 (**See color insert.**) Effect of a change of the cooperativity parameter (see different values indicated in (a) and (b)) on the opening probability of a short chain designed for loop opening. [Reprinted with permission from R. BLOSSEY and E. CARLON. Copyright (2003). American Physical Society.]

curves can be recovered with two sets of parameter (and hence, consequently, in fact by a whole range of parameter values, interpolating between the two chosen values of σ and c).

One may wonder whether the whole discussion misses an important physical parameter of DNA, the stiffness of the double helix, which we encountered before in the discussion of the *WLC model*. Since the persistence length of the double-stranded DNA, ℓ_P^{ds}, is typically much larger than that of single-stranded DNA, the relevance of excluded-volume effects between the loop and the chain has been questioned (A. HANKE and R. METZLER, 2003). In their view, the double-stranded DNA enclosing the bubble is so stiff compared to the flexible single strands of the bubble that the excluded volume effect plays no role.

Several factors, however, intervene in this issue. Firstly, denaturation is a high-temperature phenomenon, since melting temperatures typically lie in the range between 60°C - 90°C, and not at room or physiological temperature, where values of the persistence length are often considered. Secondly, in the limit of high salt concentrations, electrostatics is fully screened and the dependence of ℓ_P on temperature can be assumed to follow the worm-like chain model where $\ell_P \sim T^{-1}$.

A further effect which influences the value of the persistence length is the presence of small bubbles along the melting chain. It should be noted that the Poland-Scheraga model is tuned to describe long chains and the opening of long loops along the chain; the length of helix and loop segments should scale as the inverse cooperativity parameter, hence as $1/\sigma$. One could thus be misled to assume that short loops would be suppressed during the melting process, which is wrong.

This fact is illustrated in Figure 3.8, which plots the probability of finding a specific base pair, $i = 1$ and $i = 17$, in an open state in a chain of 30 bp length. Both $A(1)$ and $A(17)$ can be measured by fluorescence techniques; the base pair $i = 17$ is placed in the interior of an AT-rich region. The shortness of the sequence leads to the insensitivity of the melting curve on the value of c, but it sensitively depends on σ. For the small value of $\sigma \sim 10^{-4}$, both probabilities are indistinguishable, which would mean that there is no loop opening, and the sequence rather melts from the ends. Increasing the value of σ by a factor of 100 allows the opening of loops, and brings the theoretical result close to what is observed in experiment (G. ALTAN-BONNET et al., 2003). Consequently, it appears that $\sigma = \sigma(L)$.

We thus conclude that

- The Poland-Scheraga can be used to describe even complex denaturation profiles in quantitative accord with experiment;

- The theoretical results depend on the value of the critical exponent c and the cooperativity parameter σ. Both need to be (and can be) adjusted to fit to experiment;

- The fit to experiment depends on the length of the DNA under scrutiny; by construction, the Poland-Scheraga model is more tuned to DNA properties on longer length scales;

- We cannot clarify the issue of the order of the phase transition; from a practical point of view for the melting profiles of DNA this seems almost an academic question. The transition is borderline between first and second-order, beyond experimental resolution.[9]

Melting cDNA. DNA contains various sequence regions which are biological entities: first there are the *genes*, or more specifically, the protein-coding regions, the *exons*, but also the so-called junk DNA, non-coding regions, the *introns*. There are also *regulatory regions*: places where regulatory proteins attached and control the expression of genes (see Part III). If we melt purely genomic DNA, we will get a bit of everything. Can we distinguish between the melting behaviour of different sequence regions - is there a correlation between the biological purpose of a sequence and its (thermal) stability?[10] We address this question for *complementary DNA* (cDNA).

Complementary DNA (cDNA) is a molecule which contains only exons (including so-called *untranslated end regions*, UTRs). Complementary DNA can be obtained from genomic DNA by first transcribing it into RNA, splicing out the introns, and transcribing the mature RNA back into DNA; this latter step is done by a viral enzyme, the reverse transcriptase. Figure 3.9 shows the build-up of genomic DNA and cDNA in a schematic comparison.

Melting of genomic DNA of various organisms has been studied by E. YERAMIAN (2000, 2002). Quite generally it is found that the melting curves reflect the base composition; since exons are on average more GC-rich than introns, which have a bias to AT bases, this difference can in some cases permit to distinguish genes and non-coding regions from each other. AT-rich regions melt more easily since the base pairing has only two hydrogen bonds, instead of the three between G and C. However, this distinction does not always work; e.g., for bacteria it was found that the structure of their ultra-dense genomes

---

[9]Physicists protesting against this pessimistic conclusion should look back into the critical properties of classical superconductors. Further, take a look into the Additional Notes on the topic of *mixed-order transitions*.

[10]One might object that the thermal stability of DNA cannot be relevant for organisms since the melting temperatures of DNA are usually much higher than physiologically relevant temperatures. One counter-argument against this is that a position which is easier to destabilize thermally can likewise be more readily destabilized either by changes of chemical conditions (pH) or the localized action of forces, e.g., by protein complexes on DNA.

Figure 3.9 Structure of genomic DNA vs. cDNA. [Reprinted with permission from E. CARLON et al. Copyright (2005). American Physical Society.]

without non-coding regions does not permit to identify genes based on melting signatures alone.

This is indeed different for cDNA (E. CARLON et al., 2005). If one melts cDNA, one finds a differential melting curve similar to the one shown in Figure 3.10. As for genomic DNA, several distinct melting peaks arise. In the figure, sequence information has been added: the horizontal bars indicate the location of the boundaries between two exons as known from sequence annotation. The numbers on the right indicate the number of base pairs of the intron removed at that position. The vertical bars in the figure indicate when the probability of having bound pairs falls below 1/2. The ends of these bars locate the position of a *thermal boundary*: a sequence position at which the opening of the sequence is stopped at increasing temperature.

Figure 3.11 shows that in several cases thermal boundaries coincide with the annotated exon-exon boundaries, i.e., known coding regions of the genes. This observation holds for many human genes, with a resulting coincidence of thermodynamic and annotated boundaries at about 30%.

At present, there is no simple explanation for this finding; but one may speculate. In this context it is useful to think in evolutionary terms. How did exons and introns come into their positions in the first place? There are two current, opposing hypotheses on the evolution of genomes, one based on the idea that

Figure 3.10 (**See color insert.**) Melting curve and exon-exon bound-aries. For the explanation, see text; N is sequence length. [Reprinted with permission from E. CARLON et al. Copyright (2005). American Physical Society.]

exon positioning came late, and were inserted (shuffled around) into intronic DNA. (Bacteria have no introns; but according to the theory they were just very active in getting rid of the junk.) The opposing hypothesis considers exons to be evolutionarily 'early', and introns being inserted into an otherwise largely exonic DNA (W. GILBERT, 1987).

The finding that *exon-exon-boundaries* are, in humans, to about 30% located in positions at which DNA is less (thermally) stable might support the view that introns are 'late': the location of thermal boundaries might be preferred sites for intron insertion, since in these positions double-stranded DNA is more readily opened up. But then, these spots might also be prone to intron losses. In any case, an agreement between positions at which a physical signature coincides with a biological signature to such an extent seems hard to be just accidental. In an analysis of actin genes from various species cases could indeed be found which indicate an involvement of DNA denaturation physics for the insertion of introns (E. CARLON et al., 2007).

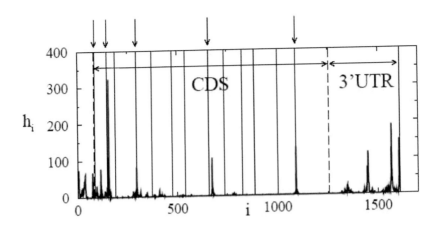

Figure 3.11 Correlation of thermal boundaries with exon-exon boundaries. [Reprinted with permission from E. CARLON et al. Copyright (2005). American Physical Society.]

DNA microarrays. The hybridization of oligomeric DNA on microarrays in order to measure gene expression levels has been a huge topic in the early days of genomics; meanwhile new techniques have largely bypassed this approach. The idea of this method is that the amount of RNA in a cell (or, of back-transcribed RNA into cDNA) can serve as a measure for the number of proteins that are translated in the cell; this is only approximately correct, but nevertheless an indication.

The recognition capability of a single-stranded RNA or DNA towards its complementary sequence can be exploited in a very simple way to measure the presence and amount of DNA in a given sample. In order to put this idea into practice, it has now become possible, using techniques from the microelectronics industry, to produce biochips - in analogy to microelectronic chips.

The principle of a *biochip*, or *DNA microarray* in our context, is the following. On a substrate, a collection of single-stranded DNA molecules, the *probes*, is built. In the case of *Affymetrix* chips, the only type of microarrays we address here, the sequences contain 25 base pairs which have been selected from a gene of interest. In fact, about 10-16 probes are produced and fixed at the substrate, reflecting different selected subsequences from the same gene. In addition, for each perfectly matching probe - one that is exactly identical to a subsequence from the selected gene - a mismatching probe is produced. This probe differs from the perfect match by one nucleotide. The mismatch probe is a control, which is used to assess *cross-hybridizations*, i.e. non-specific hybridizations.

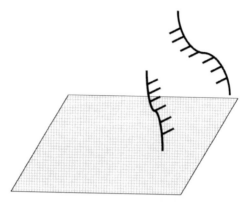

Figure 3.12 Hybridization on a microarray. Note that in the case of Affymetrix arrays, probes are DNA, while targets are RNA.

Based on this idea microarrays containing full genomes can now be built. They are used by exposing the probe strands to a set of target DNA which will bind to the selected probes, as schematically shown in Figure 3.12. The amount of bound DNA is measured, typically optically by using fluorescently labelled *target RNA*.

How do targets and probe meet each other on the chip? A very simple one is the following (E. CARLON and T. HEIM, 2005), motivated by previous work by G. A. HELD et al. (2003). After hybridizing targets and probes light intensities I of the fluorescent markers are measured. Distinguishing between specific, S, and non-specific hybridizations, N, we write

$$I(c, \Delta G) = S(c, \Delta G) + N + \epsilon \qquad (3.46)$$

where ϵ stands for purely experimental noise. I is the intensity from the probe whose complementary target RNA is present in solution at a concentration c, and ΔG is the hybridization free energy. A distinction between mismatching or matching DNA does not need to be made, since both will differ only in their values of ΔG. The non-specific hybridization contribution N depends on total RNA concentration in solution, and probably other free energy parameters reflecting partial hybridizations. The value of N is not important if one decides to consider the quantity

$$\Delta I \equiv I(c) - I(0) \approx S(c, \Delta G) \qquad (3.47)$$

Such a background subtraction is possible when we compare measurements in which c is the concentration of a particular gene; this is the case for the set of control measurements done in a *Latin square*, which consists of a well-defined dilution series for a set of specific genes added to a background, the so-called *'spike-in' genes*, and, importantly, contains the case $c = 0$.

In the simplest case, which is well applicable to chains of the order of 25 base pairs as we know from Section 3.1 of this chapter, we can understand the binding of the probe DNA and the target RNA as a two-state process, in which the target is either fully unbound in solution, or fully bound at the correct sequence at the surface. The binding is then described by a *Langmuir isotherm* of the form

$$S(c, \Delta G) = \frac{Ace^{-\beta \Delta G}}{1 + ce^{-\beta \Delta G}} \tag{3.48}$$

where, as before, $\beta = 1/RT$. The amplitude factor A sets an intensity scale; it corresponds to the saturation value in the limit $c \gg e^{\beta \Delta G}$, i.e., whenever the concentration is high or the binding energy large.

Figure 3.13 shows an example of such data analysis. Noticing from Eq. (3.48) that the free energy and the concentration appear always in the form of a product $x \equiv c \exp(-\beta \Delta G)$, we can use x as a scaling variable. The data then fall nicely on the Langmuir isotherms.

In these curves, the ΔG values were computed with parameters determined experimentally for DNA-RNA hybrids in *solution* and not for the DNA attached to the surface, and we may assume a systematic deviation between the (known) solution values and the (unknown) surface values.

This systematic deviation between the free energy values in solution and at the surface can likewise be shifted into one overall parameter: temperature. Temperature T then changes status from an experimental to a fitting parameter. If its value is chosen to lie at around 700 K, i.e., roughly twice as large as the true experimental temperature, a very good agreement between experiment and the simple theory is reached.

As seen in Figure 3.13, only the data from probe 16 deviate significantly from the computed Langmuir isotherm, but they seem to follow another isotherm which appears shifted to the right. This behaviour can be captured by introducing a probe-dependent parameter

$$\alpha_k = (1 + \tilde{c} \exp(-\beta \Delta G_{RNA}))^{-1}, \tag{3.49}$$

which takes into account the effect of RNA-RNA hybridization in solution, leading to secondary structures of the RNA strands closing in on themselves,

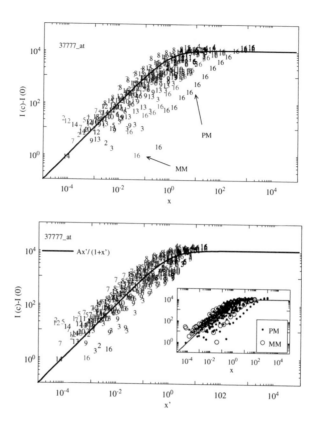

Figure 3.13 (**See color insert.**) Hybridization data falling on the Langmuir isotherm; plotted against the two scaling variables x (top) and x' (bottom) defined in the text. [Reprinted with permission from E. CARLON and T. HEIM. Copyright (2005) by Elsevier.]

thereby reducing the amount of available sequences for binding at the surface. This effect can be taken into account in the parameter combination $\tilde{c}\exp(-\beta\Delta G_{RNA})$, wherein $\tilde{c}$ is a fitting parameter. Figure 3.13 displays the full data collapse if the scaling variable

$$x' = \alpha_k c \exp -\beta\Delta G \qquad (3.50)$$

is used. What happens during hybridization in solution is indicated schematically in Figure 3.14. Sequences with very high free energies tend to hybridize to other chains present in the solution, and they are thus not available anymore

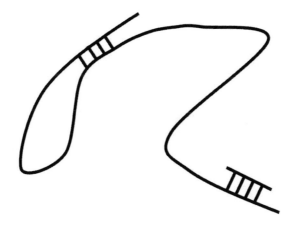

Figure 3.14 Schematic drawing of RNA hybridization in solution indicating two possible effects, self-hybridization of targets and hybridization of different fragments.

for the hybridization process at the surface - their 'active' concentration is reduced as compared to that of other sequences.

Additional Notes

RNA phase behaviour. The understanding of RNA phase behaviour has continued to expand, in particular with sophisticated renormalization group calculations (F. DAVID and K. J. WIESE, 2007 and 2009; W.D. BAEZ et al., 2018).

DNA melting. Short DNAs remain of interest due to their applicability in biotechnology, in particular in microarrays. Modifications of the linear DNA have become of interest as well; the simplest such systems are DNA hairpins, in which a single-stranded DNA bends back to itself and increases specificity, the so-called *molecular beacons* (G. BONNET et al., 1999). We will discuss some of their aspects in Part II of the book.

Other models used in studies of DNA denaturation. There are several other models that have been developed for the description of DNA denaturation aside from the Poland-Scheraga model. We here comment only on one model which follows a rather different philosophy: the Dauxois-Peyrard-Bishop model by TH. DAUXOIS et al. (1993) following earlier work by M. PEYRARD and A. R. BISHOP (1989). Their model is more microscopic than the Poland-Scheraga model; it starts out from the dynamics of the chains, i.e., the main ingredient in the model is the transverse stretching y_n of the hydrogen bonds between the complementary bases counted by the index n. The model is defined by the Hamiltonian

$$H = \sum_n \left[\frac{1}{2} m \dot{y}_n^2 + V(y_n) \right] \qquad (3.51)$$

where m is the mass of the bases in the kinetic energy. The potential model contains two contributions

$$V(y_n) = \left[D_n \left(e^{-\alpha_n y_n} - 1 \right)^2 + \frac{k}{2} \left(1 + \varrho e^{(y_n + y_{n-1})} \right) (y_n - y_{n-1})^2 \right]. \qquad (3.52)$$

The Morse potential in the first term describes the effective interactions between complementary bases: it contains both the attraction due to the hydrogen bonds forming the base pairs and the repulsion of the negatively charged phosphates in the backbone of the strands, which is screened by the surrounding solvent. The parameters D_n and α_n distinguish between the two complementary base pair combinations at site n, and hence induce a sequence dependence. The second term comprises the stacking interactions. The exponential term modifies an otherwise harmonic potential. This nonlinearity is essential: representing local constraints in nucleotide motions, it induces long-range cooperative effects. The stiffening of the coupling in the bound state compared to that in the open state leads to an abrupt entropy-driven transition. Similar conclusions were drawn by D. CULE and T. HWA (1997).

The Dauxois-Peyrard-Bishop model, as the Poland-Scheraga model, has also been confronted with experimental data, in particular for short sequences (A. CAMPA and A. GIANSANTI, 1998). Moreover, it has recently been used to correlate the dynamics of loop openings in the chain with regulatory sequence signatures. Correlations of regions with a high probability of loop openings with transcription start sites have been reported (C. H. CHOI et al., 2004) and (G. KALOSAKAS et al., 2004). More recent results are skeptical, see (T. S. VAN ERP et al., 2005).

Sequence analysis with DNA melting. The analysis of genomes by a physical analysis with DNA melting models pioneered by E. YERAMIAN has continued to find interest among researchers; however significant results are still lacking. At present, the method is best thought of enabling complementary information on DNA properties which needs to be considered with more specific biological information. A paper in this direction is by G.K. SANDVE et al., 2010. For large-scale calculations it has been demonstrated by (D. JOST and R. EVERAERS, 2009) that the use of a simpler version of the Poland-Scheraga model, the Zimm-Bragg model, is of computational advantage while being of sufficient quality for the detection of melting signals.

Mixed-order transitions and DNA melting. The DNA melting transition in the Poland-Scheraga model shows some peculiarities when compared to the standard classification of phase transitions. Classically, a first-order transition has a discontinuous order parameter, while in a second-order transition, the order parameter changes continuously, while the correlation length and the susceptibility diverge. For the PS-model, in the case of a first-order transition, the correlation length diverges algebraically, while in other models - spin models with long-range interactions - an exponential divergence is found. As we saw, for $c > 2$ the model has displayed a discontinuity of the average loop length, while the correlation length diverges as $(T - T_M)^{-1}$ at the melting transition. D. MUKAMEL and collaborators have developed a general theory of mixed-order transitions, see, e.g. (A. BAR and D. MUKAMEL, 2014) and (A. BAR et al., 2016).

DNA unzipping. The opening of DNA does not necessarily have to arise by thermal denaturation - which typically occurs at non-physiological temperatures, and hence cannot play any direct role in a living cell. Other factors, such as a change of pH or the direct application of a local force have, however, a similar effect on the molecule. These effects have been tested in single-molecule experiments, see the literature to the previous chapter.

DNA microarrays. The basic concept has been developed further in several papers by E. CARLON and his collaborators, the most recent one is by (W.W. HADIWIKARTA et al, 2012). A more general discussion of the physico-chemical basis of microarrays and other technologies is (A. HARRISON et al., 2013).

References

G. Altan-Bonnet, A. Libchaber, O. Krichevsky
Bubble dynamics in double-stranded DNA
Phys. Rev. Lett. **90**, 138101 (2003)

W.D. Baez, K.J. Wiese, R. Bundschuh
On the behavior of random RNA secondary structures near the glass transition
arXiv:1808.02351v1 (2018)

A. Bar, D. Mukamel
Mixed-order phase transition in a one-dimensional model
Phys. Rev. Lett. **112**, 015701 (2014)

A. Bar, S.N. Majumdar, G. Schehr, D. Mukamel
Exact extreme-value statistics at mixed-order transitions
Phys. Rev. E **93**, 052130 (2016)

J. Bayer, J.O. Rädler, R. Blossey
Chains, dimers, and sandwiches: melting behaviour of DNA nanoassemblies
Nano Letters **5**, 497-501 (2005)

R.D. Blake, S.G. Delcourt
Thermal stability of DNA
Nucl. Acids Res. **26**, 3323-3332 (1998)

R.D. Blake, J.W. Bizarro, J.D. Blake, G.R. Day, S.G. Delcourt, J. Knowles, K.A. Marx, J. Santa Lucia, Jr.
Statistical mechanical simulation of polymeric DNA melting with MELTSIM
Bioinformatics **15**, 370-375 (1999)

R. Blossey, E. Carlon
Reparametrizing the loop entropy weights: Effect on DNA melting curves
Phys. Rev. E **68**, 061911 (2003)

G. Bonnet, S. Tyagi, A. Libchaber, F.R. Kramer
Thermodynamic basis of the enhanced specificity of structured DNA probes
Proc. Natl. Acad. Sci. USA **96**, 6171-6176 (1999)

R. Bundschuh, T. Hwa
Statistical mechanics of secondary structures formed by random RNA sequences
Phys. Rev. E **65**, 031903 (2002)

R. Bundschuh, T. Hwa
Phases of the secondary structures of RNA sequences
Europhys. Lett. **59**, 903-909 (2002)

A. Campa, A. Giansanti
Experimental tests of the Peyrard-Bishop model applied to the melting of very short DNA chains
Phys. Rev. E **58**, 3585-3588 (1998)

E. Carlon, E. Orlandini, A. L. Stella
Role of stiffness and excluded volume in DNA denaturation
Phys. Rev. Lett. **88**, 198101 (2002)

E. Carlon, M.L. Malki, R. Blossey
Exons, introns and DNA thermodynamics
Phys. Rev. Lett. **94**, 178101 (2005)

E. Carlon, T. Heim
Thermodynamics of RNA/DNA hybridization in high density oligonucleotide microarrays
Physica A **362**, 433-449 (2005)

C.H. Choi, G. Kalosakas, K.O. Rasmussen, M. Hiromura, A.R. Bishop, A. Usheva
DNA dynamically directs its own transcription initiation
Nucl. Acids Res. **32**, 1584-1590 (2004)

E. Carlon, A. Dkhissi, M. Lejard Malki, R. Blossey
Stability domains of actin genes and genomic evolution
Phys. Rev. E **76**, 051916 (2007)

D. Cule, T. Hwa
Denaturation of heterogeneous DNA
Phys. Rev. Lett. **79**, 2375-2378 (1997)

F. David, K.J. Wiese
Systematic field theory of the RNA glass transition
Phys. Rev. Lett. **98**, 128102 (2007)

F. David, K.J. Wiese
Field theory of the RNA freezing transition
J. Stat. Mech. - Theory and Experiment, P10019 (2009)

T. Dauxois, M. Peyrard, A. R. Bishop
Entropy-driven DNA denaturation
Phys. Rev. E **47**, R44-R47 (1993)

B. Duplantier
Polymer network of fixed topology - renormalization, exact critical exponent γ in 2 dimensions, and $d = 4 - \varepsilon$
Phys. Rev. Lett. **57**, 941-944 (1986)

T.S. van Erp, S. Cuesta-Lopez, J.-G. Hagmann, M. Peyrard
Can one predict DNA transcription start sites by studying bubbles?
Phys. Rev. Lett. **95**, 218104 (2005)

M. Fixman, J.J. Freire
Theory of DNA melting curves
Biopolymers **16**, 2693-2704 (1977)

T. Garel, H. Orland
Generalized Poland-Scheraga model for DNA hybridization
Biopolymers **75**, 453-467 (2004)

W. Gilbert
The Exon Theory of Genes
vol. LII Cold Spring Harbour Symposia on Quantitative Biology (1987)

W.W. Hadiwikarta, J.-C. Walter, J. Hooyberghs, E. Carlon
Probing hybridization parameters from microarray experiments: nearest-neighbor model and beyond
Nucl. Acids Res. **40**, e138 (2012)

A. Hanke, R. Metzler
Comment on: "Why is the DNA denaturation transition first order?"
Phys. Rev. Lett. **90**, 159801 (2003)

A. Harrison et al.
Physico-chemical foundations underpinning microarray and next-generation sequencing experiments
Nucl. Acids Res. **41**, 2779-2796 (2013)

G. A. Held, G. Grinstein, Y. Tu
Modeling of DNA microarray data by using physical properties of hybridization
Proc. Natl. Acad. Sci. USA **24**, 7575-7580 (2003)

D. Jost, R. Everaers
Genome wide application of DNA melting analysis
J. Phys.: Condensed Matter **21**, 034108 (2009)

Y. Kafri, D. Mukamel, L. Peliti
Melting and unzipping of DNA
Eur. Phys. J. B. **27**, 135-146 (2002)

G. Kalosakas, K.O. Rasmussen, A.R. Bishop, C.H. Choi, A. Usheva
Sequence-specific thermal fluctuations identify start sites for DNA transcription
Europhys. Lett. **68**, 127-133 (2004)

R. Owczarzy, P.M. Vallone, F.J. Gallo, T.M. Paner, M.J. Lane, A. S. Benight
Predicting sequence-dependent melting stability of short duplex DNA oligomers
Biopolymers **44**, 217-239 (1998)

M. Peyrard, A.R. Bishop
Statistical mechanics of a nonlinear model for DNA denaturation
Phys. Rev. Lett. **62**, 2755-2759 (1989)

D. Poland
Recursion relation generation of probability profiles for specific-sequence macromolecules with long-range correlations
Biopolymers **13**, 1859-1871 (1974)

D. Poland, H.A. Scheraga
Theory of Helix-Coil Transitions in Biopolymers
Academic Press (1970)

G.K. Sandve et al.
The Genomic HyperBrowser: inferential genomics at the sequence level
Genome Biology **11**, R121 (2010)

J. Santa-Lucia Jr.
A unified view of polymer dumbbell, and oligonucleotide DNA nearest-neighbor thermodynamics
Proc. Natl. Acad. Sci. USA **95**, 1460-1465 (1998)

R. Thomas
Recherches sur la dénaturation des acides desoxyribonucléiques
Biochim. Biophys. Acta **14**, 231-240 (1954)

E. Yeramian
Genes and the physics of the DNA double-helix
Gene **255**, 139-150 (2000)

E. Yeramian
The physics of DNA and the annotation of Plasmodium falciparum
Gene **255**, 151-168 (2000)

E. Yeramian, S. Bonnefoy, G. Langsley
Physics-based gene identification: proof of concept for Plasmodium falciparum
Bioinformatics **18**, 190-193 (2002)

CHAPTER 4

Soft Matter Electrostatics

4.1 THE FREE ENERGY OF ELECTROSTATIC SYSTEMS

In this section we discuss the statistical mechanics of charged systems. These arise most typically in what are called 'soft matter systems': these are systems with some relevance to biology, like membranes, biomolecules, viruses and the like.

We start with a thermodynamic description which allows us to perform a nice exercise in Legendre transforms. In this section we follow a derivation presented in (A.C. MAGGS, 2012).

We want to discuss a system composed of a wall (or two walls) adjacent to a solution containing ions; either *counterions* that neutralize an opposite wall charge, or a salt solution. The free energy of such a system reads as

$$F = \frac{1}{2} \int d^3\mathbf{r} \int d^3\mathbf{r}' \varrho(\mathbf{r}) \frac{1}{4\pi\varepsilon_0 \mid \mathbf{r} - \mathbf{r}' \mid} \varrho(\mathbf{r}') + k_B T \int d^3\mathbf{r} \sum_j (c_j \ln(c_j - c_{j0}) - c_j)$$

(4.1)

where the total charge density is given by

$$\varrho = \sum_j \varrho_j + \varrho_f .$$

(4.2)

The last term in Eq. (4.2) is the density of fixed charges, most commonly located on the wall bounding of the system. The first term in Eq. (4.1) is the electrostatic energy according to *Coulomb's law*, while the second term is the *entropy of mixing* of the ions of type j with concentration c_j. The c_{j0} in the second term are reference concentrations, related to the chemical potential of each species via $\mu_j = -k_B T \ln c_{j0}$.

The long-ranged expression of the first term can be replaced by a local term via the introduction of the *electrostatic potential*, ϕ (*Task!*). The free energy density f then reads as

$$f = \varrho\phi - \varepsilon\frac{(\nabla\phi)^2}{2} + k_BT\sum_j(c_j\ln(c_j - c_{j0}) - c_j) \tag{4.3}$$

where ε is the *dielectric constant* at space point $\mathbf{r}$.

Looking now at the differential

$$df = d\phi(\varrho + \nabla\cdot\varepsilon\nabla\phi) + \sum_j dc_j(q_j\phi + k_BT\ln(c_j/c_{j0})) \tag{4.4}$$

we have from the second term the relation

$$q_j\phi + k_BT\ln(c_j/cj0)) = 0 \tag{4.5}$$

which when put back into f yields the expression

$$f = \varrho_f\phi - \varepsilon\frac{(\nabla\phi)^2}{2} - k_BT\sum_j c_{j0}e^{-\beta q_j\phi} \tag{4.6}$$

with $\beta = 1/k_BT$. This is the standard free energy density for the *Poisson-Boltzmann theory*, expressed in terms of the electrostatic potential. We can generalize this expression slightly by renaming the term originating from the dissociated ions as a function $g(\phi)$ which could equally well cover other specific cases. The free energy density then reads as

$$f = \varrho_f\phi - \varepsilon\frac{(\nabla\phi)^2}{2} - g(\phi). \tag{4.7}$$

There is a problem with this expression, however. The free energy according to this expression is not convex. This problem can be remedied by invoking a Legendre transform.

In order to show this we start with the electrostatic free energy density

$$u = \varrho_f\phi - \varepsilon\frac{(\nabla\phi)^2}{2} \tag{4.8}$$

and introduce the electrostatic field $\mathbf{E} = -\nabla\phi$ via a Lagrange multiplier $\mathbf{D}$ which leads to the constrained functional

$$u = \varrho_f\phi - \varepsilon\frac{\mathbf{E}^2}{2} + \mathbf{D}\cdot(\mathbf{E} + \nabla\phi). \tag{4.9}$$

Integrating this expression by parts allows to rewrite it as

$$u = -\varepsilon \frac{\mathbf{E}^2}{2} + \mathbf{D} \cdot \mathbf{E} - \phi(\nabla \cdot \mathbf{D} - \varrho_f) \,. \tag{4.10}$$

We can now vary this expression with respect to $\mathbf{E}$ and eliminate the latter via the resulting relation for the dielectric displacement field $\mathbf{D} = \varepsilon(\mathbf{r})\mathbf{E}$, yielding for u

$$u = \frac{\mathbf{D}^2}{2\varepsilon} - \phi(\nabla \cdot \mathbf{D} - \varrho_f) \,, \tag{4.11}$$

in which ϕ serves as a Lagrange multiplier to enforce *Gauss' law*.

We now go back to Eq. (4.7) and add the local contribution from the ions:

$$u = \frac{\mathbf{D}^2}{2\varepsilon} - \phi(\nabla \cdot \mathbf{D} - \varrho_f) - g(\phi) \,. \tag{4.12}$$

The minimization with respect to ϕ now corresponds to the Lagrange transform and allow to write the free energy density in the convex form

$$u = \frac{\mathbf{D}^2}{2\varepsilon} + \mathcal{L}(g)[\varrho_f - \nabla \cdot \mathbf{D}] \tag{4.13}$$

where

$$\mathcal{L}(g)[\xi] = \sup_x (x\xi - g(\xi)) \equiv \tilde{g}(\xi) \,. \tag{4.14}$$

In the case of monovalent ions one specifically has

$$g(\phi) = 2k_B T c_0 \cosh(q\beta\phi) \tag{4.15}$$

and hence

$$\tilde{g}(\xi) = \frac{k_B T \xi}{q} \sinh^{-1}(\xi/2qc_0) - k_B T \sqrt{4c_0^2 + \xi^2/q^2} \,. \tag{4.16}$$

We therefore see that the convexity problem can be easily removed by invoking a Legendre transform to a *'dual'* description in terms of the dielectric displacement field $\mathbf{D}$. The downturn is that we have traded a scalar field, the electrostatic potential ϕ, for a vector-valued field $\mathbf{D}$ with an interaction term which is generally a more complicated function than the original one.

The method works also the other way around. Suppose one starts with the free energy of a binary mixture with densities of the two components c_1 and c_2 at constant T, $f(c_1, c_2)$. The chemical potentials can be defined via

$$\mu_{1,2} = \frac{\partial f(c_1, c_2)}{\partial c_{1,2}} \tag{4.17}$$

and the pressure (or equation of state) can then be written as (see Chapter 1)

$$p(c_1, c_2) = -(f(c_1, c_2) - \mu_1 c_1 - \mu_2 c_2). \tag{4.18}$$

This equation also carries over to inhomogeneous cases in which the densities will be integrated over the occupied volume. In the case of charged particles, one generalizes

$$
\begin{aligned}
\mathcal{F}[c_1, c_2, \mathbf{D}] \;=\; & \int_V d^3\mathbf{r} \left(\frac{\mathbf{D}^2}{2\varepsilon} - \psi(\nabla \cdot \mathbf{D} - e(z_1 c_1 - z_2 c_2)) \right) \\
& - \int_V d^3\mathbf{r}\, p(\mu_1, \mu_2)
\end{aligned}
\tag{4.19}
$$

where now ψ is the Lagrange multiplier which ensures Gauss' law. Minimizing with respect to $\mathbf{D}$ one obtains the inhomogeneous thermodynamic potential

$$\mathcal{F}[\psi] = -\int_V d^3\mathbf{r} \left(\frac{1}{2}\varepsilon(\nabla\psi)^2 + p(\mu_1 - ez_1\psi, \mu_2 - ez_2\psi) \right). \tag{4.20}$$

The advantage of this expression is that it works for any equation of state p (A.C. MAGGS and R. PODGORNIK, 2016).

4.2   THE POISSON-BOLTZMANN EQUATION

The *Poisson-Boltzmann equation* can be obtained from our previous Eq. (4.6) by variation with respect to ϕ, which yields the expression

$$\nabla \cdot (\varepsilon \nabla \phi) + \sum_j q_j c_{j0} e^{-\beta q_j \phi} + \varrho_f = 0. \tag{4.21}$$

Assuming now the case of monovalent salt, $q_j = \pm q$, one has the result

$$\nabla \cdot (\varepsilon \nabla \phi) - 2qc_0 \sinh(\beta q \phi) + \varrho_f = 0. \tag{4.22}$$

Let's first look at simple (or rather, simplified) solutions of this equation. We first make clear that its behaviour is controlled by several length scales which we now introduce and discuss.

The Bjerrum length. Within our approach only two energy scales appear, the Coulomb energy between the charges and the thermal energy. Their balance allows to define the *Bjerrum length*[1]

$$\ell_B \equiv \frac{e^2}{\varepsilon k_B T} \tag{4.24}$$

which for $\varepsilon = 80$, the dielectric constant of bulk water, yields a quantitative value of $\ell_B = 7$ Å. The physical interpretation of this length scale is simple: for two oppositely charged particles at a distance $r < \ell_B$, electrostatics wins, and the particles are bound, while for $r > \ell_B$ thermal fluctuations make the particles unbind.

Debye screening length. As we see from Eq. (4.6), the Poisson-Boltzmann equation is a nonlinear equation. Under certain conditions (see below) it can be linearized, which is sometimes called the *Debye-Hückel approximation*, and reduces to

$$\nabla^2 \phi(\mathbf{x}) = \kappa^2 \phi(\mathbf{x}) \tag{4.25}$$

with

$$\kappa^{-2} = \frac{\varepsilon k_B T}{8\pi q^2 c_0} \equiv \ell_D^2 \tag{4.26}$$

where ℓ_D is the *Debye screening length*.[2] It is ion-density dependent, and varies from a value of 3 Å in 1 M NaCl to about 1 micron in pure water.

The validity of the Debye-Hückel approximation can be determined from a comparison of kinetic and interaction energies of the involved particles. If one takes $c_0^{-1/3}$ as a length describing mean particle separation, then the condition

$$\Gamma \equiv \ell_B c_0^{1/3} \ll 1 \tag{4.27}$$

needs to be fulfilled for the DH-approximation to be applicable.

---

[1]A note on units. In the discussion of electrostatics, both the CGS-system and the SI-system are used. The change between the two can be easily rationalized. Since in the discussion of electrostatics we only need the Poisson equation and little more, the translation from CGS to SI is simple. If we write Maxwell's equation in vacuum as

$$\nabla \cdot \mathbf{E} = -4\pi k \varrho \tag{4.23}$$

we get the CGS-expression with $k = 1$ and the SI-expression with $k = 1/(4\pi\epsilon_0)$. For the mathematically inclined, the choice of $k = 1/(4\pi)$ removes all units, following HEAVISIDE. Finally, the standard book on classical electrodynamics by J. D. JACKSON contains a detailed translation manual.

[2]Note that we switched to CGS units here.

Counterions at a charged planar surface: the Gouy-Chapman length.
As an exemplary application of the nonlinear Poisson-Boltzmann equation we consider the case of counterions, assumed as cations, opposite to a planar surface of negative charge. The latter is described by a surface charge density $\sigma < 0$. Since the system is translationally invariant, the Poisson-Boltzmann equation can be considered in one dimension, orthogonal to the wall in direction z, which yields, expressed in ϕ

$$\varepsilon\varepsilon_0 \phi''(z) = -4\pi\ell_B c_0 e^{-\phi(z)}. \tag{4.28}$$

This equation is complemented by the boundary condition

$$\left.\frac{d\phi}{dz}\right|_{z=0} = -\frac{4\pi}{q}\sigma > 0. \tag{4.29}$$

Upon integration one finds

$$\phi(z) = 2\ln(z + \ell_{CG}) + \phi_0 \tag{4.30}$$

where the constant potential ϕ_0 is left unspecified for now. The length ℓ_{CG} introduced here is the *Guoy-Chapman length*

$$\ell_{CG} \equiv \frac{\varepsilon k_B T}{2\pi q|\sigma|} = \frac{q}{2\pi|\sigma|\ell_B} \sim \sigma^{-1} \tag{4.31}$$

which depends on the surface charge density. The density profile of the mobile charges

$$c_0(z) = \frac{1}{2\pi\ell_B}\frac{1}{(z + \ell_{CG})^2} \tag{4.32}$$

is found to decay algebraically for large distances z, while the potential itself has a logarithmic (hence weak) divergence. The Gouy-Chapman length characterizes the charge density layer at small distances, $z \ll \ell_{CG}$.

Manning condensation. So far, our elementary reasonings allowed us to introduce a number of relevant physical length scales to gain an intuitive idea of the importance of electrostatic phenomena in solution. We now turn to a first biologically motivated application of the Poisson-Boltzmann equation. We consider the electrostatic profile of counterions around a straight cylinder of radius R. This setting can be understood as a highly idealized model for a linear, charged biomolecule like DNA or a polypeptide chain.

For this geometry, the Poisson-Boltzmann equation reads in cylindrical coordinates

$$\frac{d^2\varphi}{dr^2} + \frac{1}{r}\frac{d\varphi}{dr} + \kappa^2 e^{-\varphi(r)} = \frac{\ell_B \sigma}{Ze}\delta(r - R),\qquad(4.33)$$

where we have again transformed the potential via $\varphi = e\phi/k_B T$. This differential equation (4.33) is of *Liouville-type*, and admits an analytical solution. With the change of variable $x = R\ln(r/R)$ for $r > R$ it is transformed into

$$\frac{d^2\varphi}{dx^2} + \kappa^2 e^{-\varphi(x)+2x/R} = 0.\qquad(4.34)$$

The shifted potential $\tilde{\varphi} = \varphi(x) - 2(x/R)$ satisfies the planar Poisson-Boltzmann equation, albeit with a different boundary condition

$$\left.\frac{d\tilde{\varphi}(x)}{dx}\right|_{x=0} = \frac{\ell_B \sigma}{Ze} - \frac{2}{R}.\qquad(4.35)$$

The solution of the Poisson-Boltzmann equation is, by analogy to the planar case, given by

$$\tilde{\varphi} = 2\ln\left(1 + \frac{\kappa x}{\sqrt{2}}\right),\qquad(4.36)$$

where the boundary condition fixes the value of

$$\kappa = \frac{1}{2}\left(\frac{\ell_B \sigma}{Ze} - \frac{2}{R}\right).\qquad(4.37)$$

This result makes no sense for $\ell_B R\sigma < 2Ze$, i.e. $\kappa < 0$. We then have to take the solution of the Poisson-Boltzmann equation for $\kappa = 0$, and obtain instead a logarithmic potential

$$\varphi(r) = 2\xi_m \ln(r/R)\qquad(4.38)$$

where

$$\xi_m \equiv \frac{\ell_B R\sigma}{2Ze}\qquad(4.39)$$

is the *Manning parameter*. We thus obtain the full solution as

$$\varphi(r) = \begin{cases} 2\xi_m \ln(r/R), & \xi_m \leq 1, \\ 2\ln(r/R) + 2\ln[1 + (\xi_m - 1)\ln(r/R)], & \xi_m > 1. \end{cases}\qquad(4.40)$$

For $\xi_m > 1$, the electrostatic potential behaves like $\varphi(r) \sim 2\ln(r/R)$, and is essentially independent of charge density; the counterions are bound to the cylinder. For $\xi_m \leq 1$, the number of counterions $c_s(r) \sim exp - \phi(r)$ contained in a cylindrical shell R_0 around the cylinder behaves as

$$c_s(r) \sim r^{2(1-\xi_m)}\Big|_R^{R_0} \tag{4.41}$$

and clearly diverges with R_0: the counterions escape to infinity. The phenomenon of the counterion confinement to the DNA is called *Manning condensation* (G. S. MANNING, 1969).

4.3   PROTEIN ELECTROSTATICS

We will now go one step further and apply our knowledge to proteins. Here, the ultimate aim to determine their solvation free energies. In many cases, the dominating contribution is due to electrostatics (B. HONIG and A. NICHOLS, 1995), and the main task then is to compute the electrostatic potential of the proteins and their complexes.

The electrostatics problem associated with proteins can be determined by solving the Poisson or Poisson-Boltzmann equation for the geometry depicted in Figure 4.1. It is assumed that the space Ω is filled with an arrangement of charges (not shown explicitly), while we ignore the distribution of ions that can usually be assumed to surround the protein. We thus 'only' have to solve the Poisson equation in this inhomogeneous space. The inclusion of the mobile charges is, of course, possible.

The mathematical formulation of the problem consequently is:

$$\Delta\phi_\Omega(\mathbf{r}) = -\frac{4\pi}{\epsilon_\Omega}\varrho \tag{4.42}$$

$$\nabla_{\mathbf{r}}[\epsilon_\Sigma(r)\nabla_r\phi_\Sigma(\mathbf{r})] = 0 \tag{4.43}$$

$$\partial_n[\epsilon_\Omega\phi_\Omega - \epsilon_\Sigma\phi_\Sigma]|_\Gamma = 0 \tag{4.44}$$

$$\phi_\Omega|_\Gamma = \phi_\Sigma|_\Gamma \tag{4.45}$$

For proteins, $\epsilon_\Omega \approx 2 - 5$ is usually assumed, while for the surrounding water, $\epsilon_\Sigma \approx 80$.

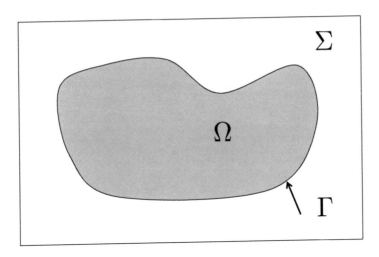

Figure 4.1 Electrostatic problem of a protein in a solvent: a body Ω imbedded in space Σ; Γ defines the boundary.

The solution of the above equations can be obtained only numerically for the complex protein surfaces. This requires the development of sophisticated solvers; this topic is discussed further in the Additional Notes. The example we show here in Figure 4.2 comes from NGUYEN et al. (2017).

Nonlocal electrostatics. In the Poisson equation for the solvent we have allowed for a (local) spatial dependence of ϵ_Σ; this is frequently done in a phenomenological way to mimic effects of the modification of the water dielectric properties near the protein. This effect is based on the restricted motion of the water near the protein, bringing about distinct *orientational polarization correlations*. We will now see how this effect of water structure can be built into the theory of electrostatics in a more fundamental way.

If we want to describe the electrostatic properties of biomolecules on length scales for which a membrane can be modeled as a fluctuating sheet, a DNA molecule as a cylinder, and a globular protein as a sphere, the level of description of electrostatic interactions is well justified. For the large protein structures we have just seen, which are cylindrical on a mesoscopic scale, but have a lot of atomistic structural detail which may - or will - be of biological relevance, we have been pushing things a bit too far. It seems hard to justify why the water surrounding a protein surface can, on this atomic scale, be considered a dielectric medium of dielectric constant $\epsilon = 80$.

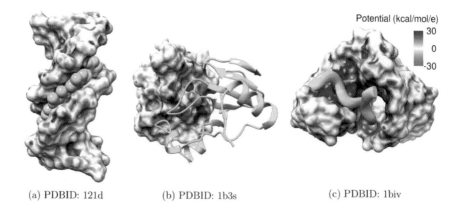

(a) PDBID: 121d   (b) PDBID: 1b3s   (c) PDBID: 1biv

Figure 4.2 (**See color insert.**) Illustration of surface electrostatic potentials (in units of $kcal/mol/e$) for three complexes, generated by Chimera software E.F. PETTERSEN et al. (2004). a) PDBID: 121d (in Drug-DNA complexes); b) PDBID: 1b3s (in barnase-barstar complexes); c) PDBID: 1biv (in RNA-peptide complexes). [Reprinted with permission by John Wiley & Sons from D.D. NGUYEN, D. WANG and G.-W. WEI (2017).]

On such length scales, the usual continuum approach ultimately must break down, since water is not featureless on these scales: it has structure. The water molecules respond to the presence of charges, and their network of hydrogen bond has to rearrange, leading to correlations of the water molecule orientations over a characteristic length scale which we denote as λ.

One way to 'repair' the error made in classical electrostatics is to modify the dielectric function. Conventionally this is done by an inclusion of a local spatial dependence $\epsilon(r)$ near the protein surface. This function, which can be parametrized in various ways, is used to effectively reduce the dielectric constant of water to a small value near the protein surface. This approach is, while physically justifiable, a technically uncontrolled procedure. Interest has therefore risen in systematic extensions of the theory of continuum electrostatics that allow to account for spatial variations of the dielectric behaviour of the solvent, in particular near the boundary of a protein.

Within the continuum theory of the electrodynamics of matter, the orientational polarizability of water around a protein is nothing but a spatial dispersion effect. Such effects are known to be tractable within electrodynamic theory, and indeed they have taken into account an approach called 'nonlocal electrostatics'[3] (A. A. KORNYSHEV et al., 1978; M. A. VOROTYNTSEV, 1978; A. A. KORNYSHEV and M. A. VOROTYNTSEV, 1979) which generalizes the commonly used electrodynamics of continuous media to take into account such spatial dispersion effects.

The basis of this theory is the linear relationship between the dielectric displacement field $\mathbf{D}$ and the electric field $\mathbf{E}$ through a permittivity kernel which, in general, depends on two spatial arguments,

$$\mathbf{D}(\mathbf{r}) = \frac{1}{4\pi} \int d\mathbf{r}' \epsilon(\mathbf{r}, \mathbf{r}') \mathbf{E}(\mathbf{r}'). \tag{4.46}$$

Here, $\epsilon(\mathbf{r}, \mathbf{r}')$ is the dielectric permittivity tensor. It carries the new characteristic length scale, the correlation length λ of the water *orientational polarization correlations* introduced before. This length defines the scale for the deviation of the dielectric properties of a solvent from its average bulk value.

For the general protein geometry of Figure 4.1, the generalization of the Poisson equation in the solvent in nonlocal electrostatics reads as

$$\nabla \int_{\Sigma} d\mathbf{r}' \epsilon_{\Sigma}(\mathbf{r}, \mathbf{r}') \nabla' \phi_{\Sigma}(\mathbf{r}') = 0 \tag{4.47}$$

where the primed symbol ∇' denotes differentiation with respect to $\mathbf{r}'$. The main ingredient of the nonlocal theory is the integral kernel $\epsilon(\mathbf{r}, \mathbf{r}')$ which contains the dependence on the correlation length λ. The mathematical expression for this model depends on the water model one wants to use. A simple, standard example for this quantity is the *Lorentzian model*[4]

$$\epsilon(|\mathbf{r} - \mathbf{r}'|) = \epsilon_{\infty} \delta(\mathbf{r} - \mathbf{r}') + \frac{\epsilon_{\Sigma} - \epsilon_{\infty}}{4\pi\lambda^2} \frac{e^{-\frac{|\mathbf{r} - \mathbf{r}'|}{\lambda}}}{|\mathbf{r} - \mathbf{r}'|}. \tag{4.48}$$

As can be observed for this example - indeed this turns out to be of rather general nature - one can write Eq. (4.48) in a more general form as

$$\epsilon(|\mathbf{r} - \mathbf{r}'|) = \epsilon_{\infty} \delta(\mathbf{r} - \mathbf{r}') + \tilde{\epsilon}\mathcal{G}(\mathbf{r} - \mathbf{r}') \tag{4.49}$$

---

[3]The name 'nonlocal electrostatics' is unfortunate since it may create confusion by alluding to strange effects of 'action at a distance' and the like. The nonlocality we talk about is nothing of that sort, it just means that the dielectric response requires to take into account field values not at local points, but over a certain spatial range. All physical fields (potentials, electrostatic fields) remain well-defined local objects.

[4]The Lorentzian model assumes isotropy of space as an additional simplifying assumption. Although this assumption does not hold strictly, it is an acceptable first approximation.

Figure 4.3 (**See color insert.**) Comparison of local (left) and nonlocal (right) electrostatic potential of the enzyme trypsin. The color code indicates positive and negative values of the potential for a selected threshold. [Reprinted with permission from Oxford University Press from A. HILDEBRANDT et al., (2006).]

where $\tilde{\epsilon} \equiv (\epsilon_\Sigma - \epsilon_\infty)/4\pi\lambda^2$, and $\mathcal{G}$ is a Green function satisfying

$$\mathcal{L}\mathcal{G} = -\delta(\mathbf{r} - \mathbf{r}') \tag{4.50}$$

with, in the given case, $\mathcal{L} \equiv \Delta - \lambda^{-2}$, and $\mathcal{G}$ as the Green function of the *Yukawa potential*. Eq. (4.49) contains two terms; the first reduces to the local limit of the dielectric function ϵ_∞ at small distances $\mathbf{r} \to \mathbf{r}'$ and is cancelled by the term $\propto \epsilon_\infty$ in the second term. The remaining contribution is the local limit for large distances, ϵ_Σ, the usual macroscopic dielectric constant.

Eq. (4.49) can be taken as the starting point to reformulate the theory of nonlocal electrostatics as a local theory. For this, we in addition represent the dielectric displacement field as a sum of an irrotational part and a solenoidal part (a so-called *Helmholtz decomposition*),

$$\mathbf{D}(\mathbf{r}) = -\nabla\psi(\mathbf{r}) + \nabla \times \xi(\mathbf{r}). \tag{4.51}$$

One can show that the solenoidal part, although it explicitly appears in the expression of the dielectric displacement field, does not appear in the equations of the electrostatic potentials (A. HILDEBRANDT, 2005).

We are then finally left with the following system of equations for the electrostatic potentials ϕ and ψ, (A. HILDEBRANDT et al., 2004)

$$\Delta\phi_\Omega = -\frac{4\pi}{\epsilon_\Omega}\varrho \tag{4.52}$$

$$\Delta\psi_\Sigma = 0 \tag{4.53}$$

$$\epsilon_\Omega\, \partial_n\phi_\Omega|_\Gamma = \partial_n\psi_\Sigma|_\Gamma \tag{4.54}$$

$$\phi_\Omega|_\Gamma = \phi_\Sigma|_\Gamma \tag{4.55}$$

$$[\epsilon_\infty\mathcal{L} - \tilde{\epsilon}]\,\phi_\Sigma = 4\pi\mathcal{L}\psi_\Sigma(\mathbf{r})\,. \tag{4.56}$$

Within its reformulation in terms of the local fields ϕ and ψ, the theory of nonlocal electrostatics can now be treated with standard approaches in order to numerically solve them with boundary element methods.

Such a result is shown in Figure 4.3. In this figure, the local and the nonlocal electrostatic potentials of the enzyme trypsin are shown for comparison. It can clearly be seen that the structure of the electrostatic potential obtained in the nonlocal description deviates significantly from the local one which, for the same threshold value of the potential surfaces, hardly reaches beyond the geometric structure of the protein surface.

From this application we can deduce that the electrostatic potential of proteins on Ångstrom scales is markedly influenced by the water properties, and that electrostatic effects are important for the 'visibility' of a protein to its interaction partners in solution.

4.4 CHROMATIN ELECTROSTATICS

Chromatin structure. In this section we break out of the central dogma of molecular biology, DNA → RNA → protein, which has so far been our guiding principle. In any organism all these molecules have to act in concert, and the linear sequence should thus be better represented by a diagram with multiple feedback loops.

In order to express genes, they first have to be switched on. This is done in part by specific proteins, the *transcription factors* in prokaryotes - and rather *transcription factories*, in eukaryotes. The function of these molecules requires recognition capability: proteins can detect their binding sites with high specificity. We will come back to this in Part II of the book when we look at the dynamics of transcription factor binding on DNA.

In eukaryotes an additional problem arises, however. DNA is condensed and packed in the cellular nucleus, making a simple DNA-protein recognition process based on a diffusional search impossible. There must be a way for a protein or enzyme to find out where it has to go in the first place, even when the DNA is still condensed. Or, DNA has to be unpacked only partially, such that the search time for transcription factors becomes reasonably short.

In order to understand the regulation of transcription in eukaryotes, we first have to get an idea of the compact form of DNA in the cell. The condensed form of DNA, in which it is actually a highly organized DNA-protein complex, is called *chromatin*. Chromatin is the structural basis of chromosomes. The different levels of organisation which are at present only partially understood. Our current knowledge is summarized in Figure 4.4 (from G. FELSENFELD and M. GROUDINE, 2003).

The best understood element of chromatin structure is the *nucleosome*. A nucleosome is a DNA-protein complex which consists of an octameric protein complex built out of eight *histone proteins* (of four different kinds), around which DNA is wrapped along 146 bp with 1.75 left-winded turns. A *linker histone* at the exit of the DNA histones completes the structure; the precise positioning of the linker histone is not yet known. The molecular structure of the so-called *nucleosomal core particle* - the DNA wrapped around the histone octamer - has meanwhile been spatially resolved down to 1.9 Å. A ribbon representation of a nucleosomal particle is shown in Figure 4.5.

Unfortunately, the crystal structure of a nucleosome does not tell its full story. The N-terminal tails of the histone proteins are largely unstructured random coils, hence they do not crystallize. These tails are functionally important in two ways:

- The tails are positively charged and can interact with the negatively charged DNA to form the condensed structures of chromatin, leading to both local or global structural changes of chromatin structure;

- The tails are subject to various chemical modifications, brought about by specific enzymes. Some of these modifications can alter the charge of the chain. At the same time, many of the modifications can be read

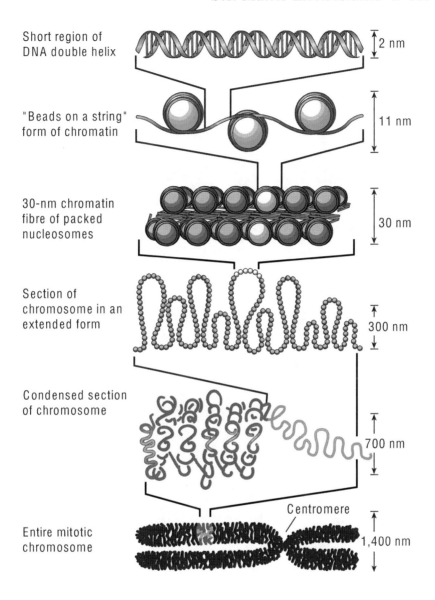

Short region of
DNA double helix

2 nm

"Beads on a string"
form of chromatin

11 nm

30-nm chromatin
fibre of packed
nucleosomes

30 nm

Section of
chromosome in an
extended form

300 nm

Condensed section
of chromosome

700 nm

Centromere

Entire mitotic
chromosome

1,400 nm

Figure 4.4 (**See color insert.**) Chromatin structure. The base level of compaction is the 11 nm fibre, in which the nucleosomes are positioned as 'beads on a string'; nucleosomal arrays condense to form the 30 nm fibre of yet unknown structure; higher order compactions lead to the familiar chromosome shape of condensed chromatin. [Reprinted with permission from Springer Nature from G. FELSENFELD and M. GROUDINE (2003).]

Figure 4.5 (**See color insert.**) Nucleosomal core particle in a ribbon presentation of the core histones bound to DNA (blue). Notice the dangling histone tails. Generated with Chimera from the nucleosome structure PDB 1KX5.

```
         p     ac
H2A  N - S G R G K Q G G K A R A K A K T R S S R A G L
         5         10      15      20

      rib   ac            ac  ac          ac                     p
H2B  N - P E P S K S A P A P K K G S K K A I T K A Q K K D G K K R K R S R K
         5         10      15      20      25      30

              m        m/ac p     ac    ac        ac      m p              m
H3   N - A R T K Q T A R K   S T G G K A P R K Q L A T K A A R K S A P A T G G V K K
             5         10      15      20      25      30      35

       p     ac  ac      ac    ac      m
H4   N - S G R G K G G K G L G K G G A K R H R K V L R D
         5         10      15      20
```

Figure 4.6 Specific enzymatic modifications on histone tails. (Rib = ribosylation, a rare modification.)

by specific proteins which affect chromatin structure locally. It has been argued that the histone modifications constitute a higher level regulatory code (B. D. STRAHL and C. D. ALLIS, 2000).

These two mechanisms are clearly not fully independent from each other, and are generally referred to as particular examples of *chromatin remodelling*. The notion of remodelling stands for the totality of dynamic structural changes chromatin can undergo and which are relevant for the regulation of transcription.

Histone tail modifications. Enzymatic histone tail modifications are amino acid specific: not only the chemical nature of the amino acid is a determinant, but also its position on the tail. This is indicated in Figure 4.6. Several different types of modifications are presently known. The most important are:

- *Acetylation.* One or several acetyl groups (CH_3CO) are transferred to lysine residues; the addition of one lysine reduces the positive charge of the tail by one. The modification is reversible, i.e., acetylating and deacetylating enzymes have been found.

- *Methylation.* One (or more) methyl groups (CH_3) are added to lysine residues; there is no effect on the charge. The reversibility of this modification is still under scrutiny; the first demethylating enzymes have recently been found, as discussed in A. BANNISTER and T. KOUZARIDES (2005).

- *Phosphorylation.* This modification can affect several residues: serine, threonine, tyrosine on the one hand, and lysine, histidine, and arginine on the other. The modifications are chemically different for both groups of amino acids.

In general, histone tail modifications do differ somewhat between organisms, despite the high evolutionary conservation of the histones, their tails included. While the experimental evidence is growing that the histone tail modifications are read by a specific transcription machinery (TH. AGALIOTI, G. CHEN and D. THANOS, 2002), and apparently different modifications do not act independently (W. FISCHLE, Y. WANG and C. D. ALLIS, 2003), it is clear that the underlying 'background' mechanism is based on an electrostatic attraction of the negatively charged DNA and the positively charged histones. It is thus of interest to devise a simplified model view of this compaction mechanism.

Chromatin fibre electrostatics. We here follow H. SCHIESSEL (2002). He considered the electrostatic interaction between two DNA strands at the entry-exit point of the DNA wrapped around the nucleosome, see Figure 4.7. The

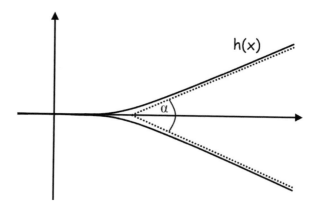

Figure 4.7 Schematic drawing of the entry-exit region around a nucleosome. The two bent lines represent the lower and upper DNA strands; the entry-exit angle α is defined via the indicated asymptotics.

angle α between the two DNA strands defines the *entry-exit angle* at the nucleosome. For the geometry of Figure 4.7, it is defined by

$$h'(\infty) \equiv \tan(\alpha/2). \tag{4.57}$$

The relevant question to be answered is that of the dependence of the angle α on salt concentration c_s, or on other charge-affected quantities. Salt contributions are relevant for the screening of electrostatic interactions, since the Debye screening length behaves as

$$\kappa^{-1} = \ell_D = (8\pi c_s \ell_B)^{-1/2} \tag{4.58}$$

where ℓ_B is the *Bjerrum length*, the measure for the respective importance of electrostatic forces and thermal energy we had introduced before in Eq. (4.24).

The DNA strands are considered as two semi-flexible polymers at positions $\pm h(x)$ with persistence length ℓ_P and a *line-charge density* λ. Their free energy is approximately given by

$$F \approx 2k_b T \int_0^\infty dx \left[\frac{\ell_P}{2} (h'')^2 + \ell_B \lambda^2 K_0(2\kappa h(x)) \right] \qquad (4.59)$$

where $K_0(x)$ is the *modified Bessel function* of zero order.[5] The first term in this expression is the bending energy of the strands, while the second describes their electrostatic interaction. The form of this term applies if it can be assumed that the interaction can be considered as that of a straight chain (hence the function K_0) and a single charge at distance $2h$. This approximation works as long as α is not too large; it tends to underestimate the true value.

The Euler-Lagrange equation for $h(x)$ reads, using the property $K_0'(x) = -K_1(x)$ of the Bessel function,

$$\ell_P h'''' = 2\ell_B \lambda^2 \kappa K_1(2\kappa h) \qquad (4.60)$$

together with the boundary conditions

$$h(0) = h'(0) = 0 = h''(\infty) = h'''(\infty). \qquad (4.61)$$

A variable change $\tilde{h} = 2\kappa h$, $\tilde{x} = (4l_B \lambda^2 \kappa^2)/l_P)^{1/4} x$ allows to express Eq. (4.60) as $h'''' = K_1(h)$ (dropping tildes). With the relation $s = h'(x)|_{x=\infty}$ one immediately obtains the dependence of the opening angle α on the physical parameters, in particular the length scales we defined in Section 4.3,

$$\tan(\alpha/2) = \frac{s}{\sqrt{2}} \left(\frac{\ell_B}{\ell_P} \right)^{1/4} \left(\frac{\lambda}{\kappa} \right)^{1/2}. \qquad (4.62)$$

The value of the constant s in this expression is of $O(1)$ (H. Schiessel, 2002).

---

[5] K_α is defined by

$$K_\alpha(x) = \frac{\pi}{2} i^{\alpha+1} H_\alpha^{(1)}(ix)$$

where

$$H_\alpha^{(1)}(x) = J_\alpha(x) + iY_\alpha(x).$$

J_α is the *Bessel function* we have introduced before; the *Neumann function* $Y_\alpha(x)$ is defined by

$$Y_\alpha(x) = \frac{J_\alpha(x) \cos \alpha\pi - J_{-\alpha}(x)}{\sin \alpha x}.$$

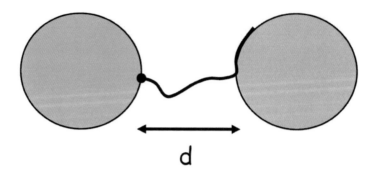

Figure 4.8 Idealization of tail bridging: a peptide chain connects two spherical particles.

Solving for α, one finds the dependence

$$\alpha = \arctan(C\, c_s^{-1/4}). \qquad (4.63)$$

Intramolecular electrostatic effects can also be included in the calculation. In this case one has to consider the modification of the persistence length ℓ_P by the charges along the chain: due to their mutual repulsion the charges along the molecule increase the rigidity of the chain. This additional repulsion can be calculated and leads to a modification of the persistence length ℓ_P, which has to be replaced by the *Odijk-Skolnick-Fixman persistence length*,

$$\ell = \ell_P + \ell_{OSF} = \ell_P + \ell_B \lambda^2/(4\kappa^2). \qquad (4.64)$$

Plugging in numbers into the result Eq. (4.63), one finds α-values of about $51°$ for 15 mM salt, and a value of $64°$ for 5 mM, in line with the expectation that a decrease of salt concentration favours the opening of the structure.

The value of the entry-exit angle can be influenced by the histone tails if they bind to the DNA. This is a local effect, when the tails bind to the DNA attached to its own histone octamer.

There is, however, also the important possibility that the tails of different nucleosomes interact. This gives rise to a *tail-bridging effect* (F. MÜHLBACHER et al. 2006), which is illustrated in Figure 4.8.

Assuming that the nucleosomes can, in a zeroth-order approximation, as spherical particles of radius R, the interaction potential between them is given by a contribution of the repulsion between the two like-charge particles and the attraction caused by a bridge formed by a tail of contour length ℓ with the line density λ,

$$U = U_{rep} + U_{bridge} \tag{4.65}$$

which is explicitly given by (ignoring precise numerical prefactors)

$$
\begin{aligned}
\frac{U(d)}{k_B T} = {} & \frac{\ell_B Z^2}{(1 + R/\ell_D)^2} \frac{\exp{-d/\ell_D}}{d + 2R} \\
& - \exp\left(-(d - \ell_D)\frac{\ell_B \ell_D Z\lambda}{R^2}\right) + \exp\left(-(\ell - \ell_D)\frac{\ell_B \ell_D Z\lambda}{R^2}\right).
\end{aligned}
\tag{4.66}
$$

Based on this result one can study how the minimum of the potential changes as the line charge is varied. The minimum distance of the two particles, $d_{min} \sim \ln(C \cdot \lambda)$, where C is a constant given by the other parameters of the potential. The potential minimum becomes very shallow for decreasing λ, and finally vanishes.

Although these considerations give an idea about how electrostatic effects of the histone tails influence chromatin structure, there is still a large disparity between the understanding of any *specific* histone tail modification and the histone code hypothesis on the one side, and the physical compaction mechanism on the other. Given the highly unspecific nature of the electrostatic interactions, and the apparent high specificity with which particular tail modifications can be placed and read, a satisfactory understanding of transcriptional regulation of chromatin seems still far out.

In particular, it is yet open whether the histone tail specifications, highly specific as they may be, really constitute a code. Recent experimental work by M. F. DION et al., 2005, on the acetylation of histone H4 provides evidence that only a very restricted number of these histone tail modifications are indeed 'coding' for specific regulatory processes. The conclusions support earlier work by Q. REN and M. A. GOROVSKY, 2001, whose analysis of the acetylation patterns of a histone variant H2A.Z demonstrated the dominant role of rather non-specific electrostatic effects.

Polyelectrolytes and polyampholytes. In the discussion of the histone tails, we have not treated the full statistics of the chain itself. We close this section with generalizing the Poisson-Boltzmann equation to the case of *polyelectrolytes* - polymers like negatively charged DNA - or *polyampholytes* - polymers like the histone tails carrying both negative and positive charges. This topic has become of renewed interest recently due to the experimental investigations of intracellular phase separation, see Additional Notes. We will not go so far here, but only show the simplest way how to introduce the combination of the electrostatic degrees of freedom coupled to the density of a polyelectrolyte.

We consider a *smeared* polyelectrolyte. If the polymer carries charges, we assume that the total charge of the chain is equally distributed with each monomer carrying a fractional charge pe, where e is the electron charge. One can then write down a partition function (BORUKHOV et al., 1998)

$$Z = \int D[\psi]D[\phi]\exp(-\beta F[\psi, \phi]) \qquad (4.67)$$

in which ψ is the electrostatic potential, and ϕ the polymer monomer density. The expression of the free energy functional $F[\psi, \phi]$ is given by

$$\begin{aligned}\beta F[\psi, \phi] &= \int d\mathbf{r}(-\beta\frac{\varepsilon}{8\pi}(\nabla\psi)^2 + \sum_{\pm} c_b^{\pm}(1 - e^{\mp\beta e\psi(\mathbf{r})}) \\ &+ \frac{a^2}{6}(\nabla\phi)^2 + \frac{v}{2}\phi^4(\mathbf{r}) - \mu_p\phi^2(\mathbf{r}) + \beta pe\psi(\mathbf{r})\phi^2(\mathbf{r})). \end{aligned} \qquad (4.68)$$

In this expression, the free energy density of the monomers is parametrized by a and v, the latter being the excluded volume interaction, and the chemical potential μ_p. One notes the coupling term between the electrostatic potential $\psi(\mathbf{r})$a and the density field $\phi(\mathbf{r})$.

Upon variation with respect to the fields the mean-field equations follow; first the Poisson-Boltzmann equation

$$\varepsilon\Delta\psi(\mathbf{r}) = 8\pi c_b \sinh(\beta e\psi) - 4\pi e(p\phi^2 - p\phi_b^2 e^{\beta e\psi}), \qquad (4.69)$$

where the last term stems from the counterions dissociated from the chains. The condition of charge neutrality has been imposed: $c_b^+ \to c_b$ and $c_b^- \to c_b + p\phi_b^2$. It follows the equation for $\phi(\mathbf{r})$,

$$\frac{a^2}{6}\Delta\phi(\mathbf{r}) = v(\phi^3 - \phi_b^2\phi) + \beta pe\psi\phi. \qquad (4.70)$$

In this equation, the chemical potential μ_p has been substituted to identify the proper bulk limit, $\psi \to 0$ and $\phi^2 \to \phi_b^2$.

We will not further discuss the properties of these equations, as they can only be solved numerically; for details, see BORUKHOV et al., 1998). The point we want to make here is that the extension of the Poisson-Boltzmann approach to situations in which the statistics of the charged objects themselves comes into play - rather than just considering fixed charge distributions - rapidly leads to problems that withstand analytic treatment. They are of immense biological relevance: e.g. if one considers such polyelectrolytes confined to shells, one has a simple yet realistic model for a DNA or RNA in a virus. Some literature suggestions in this direction can be found in the Additional Notes.

Finally, we have not discussed any fluctuation effects in this section, as this demands a technical level which is borderline to the ambitions of this book. The interested reader is asked to consult the Additional Notes for suggestions where to continue in this direction from the basis that has been laid here.

Additional Notes

Theory of soft matter electrostatics. Poisson-Boltzmann theory as it has been developed here is valid only for small surface charges, and when fluctuation and correlation effects can be neglected. The latter e.g. arise in the presence of walls, when ions start to interact with their image charges - these are not taken into account for by the theory. Things get really bad when the charges have high valencies; then the whole description breaks down. The problem of going 'beyond' Poisson-Boltzmann theory has been very active in the last years. A brief account of the problems encountered there is by (A. Naji et al., 2013).

Treatment of fluctuations within Poisson-Boltzmann theory. Two standard approaches are the inclusion of one-loop corrections around the mean-field theory, and a variational approach. Ground laying papers are by R.R. Netz and H. Orland (2000, 2003). The variational approach has seen significant development towards applications, see e.g. (S. Buyukdagli and R. Blossey, 2016).

Dual Poisson-Boltzmann. The dual approach was developed by (A.C. Maggs, 2012) and has so far seen some formal development. It can presumably become very useful for complex systems, as the corresponding partition function can be evaluated by numerical minimization methods due to the inherent convexity of the functional. It also allows for the inclusions of fluctuations, see (R. Blossey and A.C. Maggs, 2018).

Manning condensation. The Manning condensation problem is a great classic in the field. It has indeed aspects of a true phase transition which were discussed in some detail in the literature. Exemplary references are (B. O'Shaughnessy and Q. Yang, 2005; A.Naji and R.R. Netz, 2005 and 2006; M. Cha et al., 2017). A more basic introductory paper on the nonlinear screening of charged macromolecules is by (G. Tellez, 2011).

Protein electrostatics based on the Poisson-Boltzmann equation. The development of Poisson-Boltzmann solvers for the calculation of protein electrostatics has developed into a large field; for a recent review, see (C. Li at al., 2013). A pioneering paper in this field is by (N.A. Baker et al., 2001).

Going beyond Poisson-Boltzmann: inclusion of water structure in the continuum theory. One strategy to improve on Poisson-Boltzmann theory for applications to biomolecules has been to include effective water properties; one such approach - phenomenological, 'nonlocal' electrostatics approach - has been pursued further by several researchers. Relevant references are (J.P. Bardhan at al., 2015; D. Xie and Y. Jiang, 2016).

An alternative approach has been to include water molecules in an explicit fashion into the continuum theory. A prominent model in this line of research has been the Dipolar Poisson-Boltzmann Langevin model (DPBL) in which water molecules are described by point-dipoles. The approach is developed in (P. KOEHL et al., 2009).

A review putting the above described and other further approaches into perspective is by (J.P. BARDHAN, 2012).

Polyampholytes. Polyampholytes are of particular current interest as *intrinsically disordered proteins* fall into their class. A variational theory for charge-disordered polymers was developed in (K. SHEN and Z.-G. WANG, 2018). A review on the biological relevance is by D.M. MITREA and R.W. KRIWACKI (2016).

References

T. Agalioti, G. Chen, D. Thanos
Deciphering the Transcriptional Histone Acetylation Code for a Human Gene
Cell **111**, 381-392 (2002)

N.A. Baker, D. Sept, S. Joseph, M.J. Holst, J.A. McCammon
Electrostatics of nanosystems: application to microtubules and the ribosome
Proc. Natl. Acad. Sci. USA **98**, 10037-10041 (2001)

A. Bannister, T. Kouzarides
Reversing histone methylation
Nature **436**, 1103-1106 (2005)

J.P. Bardhan
Biomolecular electrostatics - I want your solvation (model)
Computational Science & Discovery **5**, 013001 (2012)

J.P. Bardhan, M.G. Knepley, P. Brune
Nonlocal electrostatics in spherical geometries using eigenfunction expansions of boundary-integral operators
Mol. Based Math. Biol. **3**, 1-22 (2015)

R. Blossey, A. C. Maggs
A fluctuation-corrected functional of convex Poisson–Boltzmann theory
J. Phys. A: Math. Theor. **51**, 385001 (2018)

I. Borukhov, D. Andelman, H. Orland
Random polyelectrolytes and polyampholytes in solution
Eur. Phys. J. B **5**, 869-880 (1998)

S. Buyukdagli, R. Blossey
Beyond Poisson-Boltzmann: fluctuations and fluid structure in a self-consistent theory
J. Phys.: Condens. Matter **28**, 343001 (2016)

M. Cha, J. Yi, Y. Woon Kim
Scaling and criticality of the Manning transition
Eur. Phys. J. E **40**, 70 (2017)

M. Cha, J. Yi, Y. Woon Kim
Hidden criticality of counterion condensation near a charged cylinder
Sci. Rep. **7**, 10551 (2017)

M. F. Dion, S. J. Altshuler, L. F. Wu, O. J. Rando
Genomic characterization reveals a simple histone H4 acetylation code
Proc. Natl. Acad. Sci. USA **102**, 5501-5506 (2005)

G. Felsenfeld, M. Groudine
Controlling the double helix
Nature **421**, 448-453 (2003)

W. Fischle, Y. Wang, C. D. Allis
Binary switches and modification cassettes in histone biology and beyond
Nature **425**, 475-479 (2003)

A. Heifetz, E. Katchalski-Katzir, M. Eisenstein
Electrostatics in protein-protein docking
Protein Science **11**, 571-587 (2002)

A. Hildebrandt, R. Blossey, S. Rjasanow, O. Kohlbacher, H.-P. Lenhof
Novel formulation of nonlocal eletrostatics
Phys. Rev. Lett. **93**, 108104 (2004)

A. Hildebrandt
Biomolecules in a structured solvent
Rhombos-Verlag (2005)

A. Hildebrandt, R. Blossey, S. Rjasanow, O. Kohlbacher, H.-P. Lenhof
Electrostatic potentials of proteins in water: a structured continuum approach
Bioinformatics **23**, e99-e103 (2006)

B. Honig, A. Nichols
Classical electrostatics in biology and chemistry
Science **268**, 1144-1149 (1995)

J.D. Jackson
Classical Electrodynamics
3rd ed., Wiley (1999)

R.M. Jackson, M.J.E. Sternberg
A continuum model for protein-protein interactions: application to the docking problem
J. Mol. Biol. **250**, 258-275 (1995)

P. Koehl, H. Orland, M. Delarue
Beyond the Poisson-Boltzmann model: modeling biomolecule-water and water-water interactions
Phys. Rev. Lett. **102**, 087801 (2009)

A.A. Kornyshev, A.I. Rubinstein, M.A. Vorotyntsev
Model nonlocal electrostatics: I
J. Phys. C: Solid State Phys. **11**, 3307-3322 (1978)

A.A. Kornyshev, M.A. Vorotyntsev
Model nonlocal electrostatics: III. Cylindrical interface
J. Phys. C: Solid State Phys. **12**, 4939-4946 (1979)

C. Li, L. Li, M. Petukh, E. Alexov
Progress in developing Poisson-Boltzmann equation solvers
Mol. Based Math. Biol. **1**, 1 (2013)

A.C. Maggs
A minimizing principle for the Poisson-Boltzmann equation
Europhys. Lett. **98**, 16012 (2012)

A.C. Maggs, R. Podgornik
General theory of asymmetric steric interactions in electrostatic double layers
Soft Matter **12**, 1219-1229 (2016)

G.S. Manning
Limiting laws and counterion condensation in polyelectrolyte solutions I.
J. Chem. Phys. **51**, 924-933 (1969)

D.M. Mitrea, R.W. Kriwacki
Phase separation in biology; functional organization of a higher order
Cell Communication and Signaling, **14:1** (2016)

F. Mühlbacher, C. Holm, H. Schiessel
Controlled DNA compaction within chromatin: the tail-bridging effect
Europhys. Lett. **73**, 135-141 (2006)

A. Naji, R.R. Netz
Counterions at charged cylinders: criticality and universality beyond mean-field theory
Phys. Rev. Lett. **95**, 185703 (2005)

A. Naji, R.R. Netz
Scaling and universality in the counterion-condensation transition at charged cylinders
Phys. Rev. E **73**, 056105 (2006)

A. Naji, M. Kanduc, J. Forsman, R. Podgornik
Perspective: Coulomb fluids: weak coupling, strong coupling, in between and beyond
J. Chem. Phys. **139**, 150901 (2013)

R.R Netz, H. Orland
Beyond Poisson-Boltzmann: fluctuation effects and correlation functions
Eur. Phys. J. E **1**, 203-214 (2000)

R.R Netz, H. Orland
Variational charge renormalization in charged systems
Eur. Phys. J. E 11, 301-311 (2003)

D.D. Nguyen, B. Wang, G.-W. Wei
Accurate, robust, and reliable calculations of Poisson-Boltzmann binding energies
J. Comp. Chem. **38**, 941-948 (2017)

B. O'Shaughnessy, Q. Yang
Manning-Oosawa counterion condensation
Phys. Rev. Lett. **94**, 048302 (2005)

E.F. Pettersen, T.D. Goddard, C.C Huang, G.S. Couch, D.M. Greenblatt, E.C. Meng, T.E. Ferrin
UCSF Chimera - a visualization system for exploratory research and analysis
J. Comput. Chem. **25**, 1605-1612 (2004)

Q. Ren, M. A. Gorovsky
Histone H2A.Z Acetylation Modulates an Essential Charge Patch
Molecular Cell **7**, 1329-1335 (2001)

H. Schiessel
How short-ranged electrostatics controls the chromatin structure on much larger scales
Europhys. Lett. **58**, 140-146 (2002)

K. Shen, Z.-G. Wang
Polyelectrolyte chain Structure and solution phase behavior
Macromolecules **51**, 1706-1717 (2018)

T. Simonson
Macromolecular electrostatics: continuum models and their growing pains
Curr. Op. Struct. Biol. **11**, 243-252 (2001)

B. D. Strahl and C. D. Allis
The language of covalent histone modifications
Nature **403**, 41-45 (2000)

G. Tellez
Nonlinear screening of charged macromolecules
Phil. Trans. R. Soc. A **369**, 322-334 (2011)

M.A. Vorotyntsev
Model nonlocal electrostatics: II, Spherical interface
J. Phys. C: Solid State Phys. **11**, 3323-3331 (1978)

D. Xie, Y. Jiang
A nonlocal modified Poisson-Boltzmann equation and finite element solver for computing electrostatics of biomolecules
J. Comp. Phys. **322**, 1-20 (2016)

Figure 2.9 Graphical representation of RNA with pseudoknots (see text).

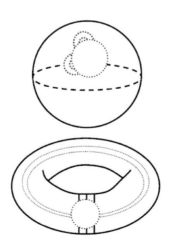

Figure 2.10 Drawing RNA on a sphere and on a torus.

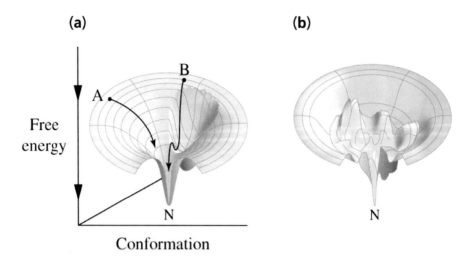

Figure 2.13 The folding funnel; a) folding paths from two different initial conditions: A follows a direct path whereas B is trapped in a metastable intermediate; b) rugged energy landscape with multiple minima at intermediate energies. Note that the 'true' equilibrium has a much lower free energy. [Reprinted with permission from John Wiley & Sons (H.S. CHAN and K.A. DILL, 1998).]

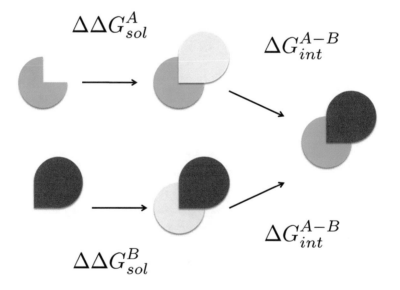

Figure 2.14 Solvation of a protein complex in water; for explanation, see text.

Figure 2.17 Force-extension curve for a single-molecule DNA. (Reprinted from C. BUSTAMANTE et al., 2000, with permission from Elsevier.)

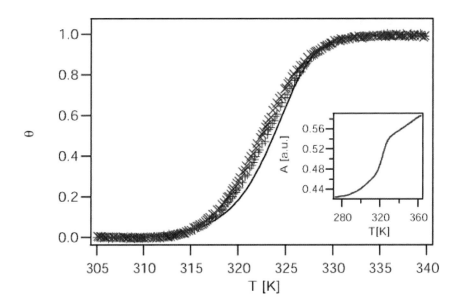

Figure 3.1 Melting curve of a 16-bp oligomer; comparison of theory and experiment. [Reprinted with permission from J. BAYER et al. Copyright (2005) American Chemical Society.]

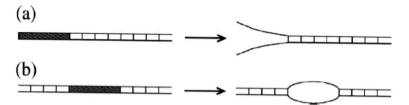

Figure 3.4 Schematic diagram of the linearization of an AT-insertion in a GC-rich bacterial plasmid. The bar represents the inserted DNA segment which upon cutting is either placed at an extremity of the molecule (a) or at its center (b). [Reprinted with permission from R. BLOSSEY and E. CARLON. Copyright (2003). American Physical Society.]

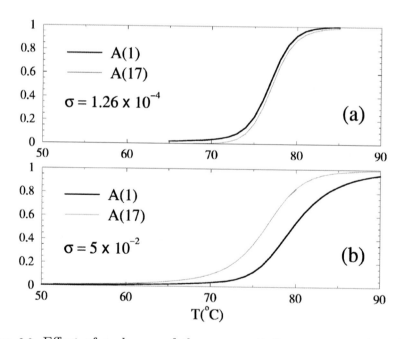

Figure 3.8 Effect of a change of the cooperativity parameter (see different values indicated in (a) and (b)) on the opening probability of a short chain designed for loop opening. [Reprinted with permission from R. BLOSSEY and E. CARLON. Copyright (2003). American Physical Society.]

Figure 3.10 Melting curve and exon-exon boundaries. For the explanation, see text; N is sequence length. [Reprinted with permission from E. CARLON et al. Copyright (2005). American Physical Society.]

Figure 3.13 Hybridization data falling on the Langmuir isotherm; plotted against the two scaling variables x (top) and x' (bottom) defined in the text. [Reprinted with permission from E. CARLON and T. HEIM. Copyright (2005) by Elsevier.]

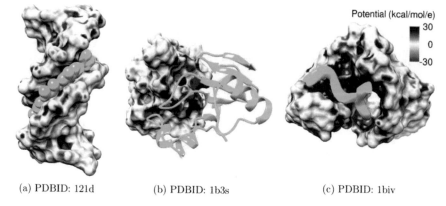

Potential (kcal/mol/e)
30
0
-30

(a) PDBID: 121d   (b) PDBID: 1b3s   (c) PDBID: 1biv

Figure 4.2 Illustration of surface electrostatic potentials (in units of $kcal/mol/e$) for three complexes, generated by Chimera software E.F. PETTERSEN et al. (2004). a) PDBID: 121d (in Drug-DNA complexes); b) PDBID: 1b3s (in barnase-barstar complexes); c) PDBID: 1biv (in RNA-peptide complexes). [Reprinted with permission by John Wiley & Sons from D.D. NGUYEN, D. WANG and G.-W. WEI (2017).]

Figure 4.3 Comparison of local (left) and nonlocal (right) electrostatic potential of the enzyme trypsin. The color code indicates positive and negative values of the potential for a selected threshold. [Reprinted with permission from Oxford University Press from A. HILDEBRANDT et al., (2006).]

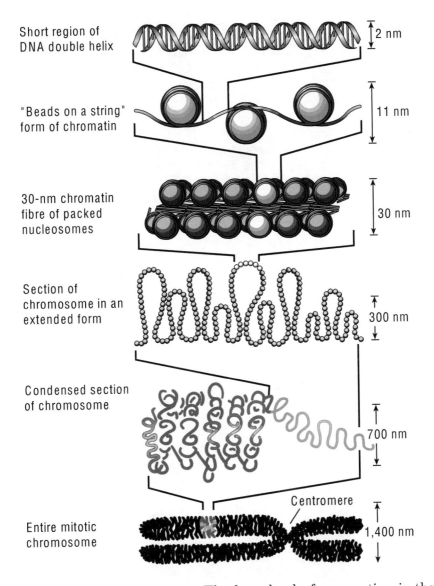

Short region of
DNA double helix

2 nm

"Beads on a string"
form of chromatin

11 nm

30-nm chromatin
fibre of packed
nucleosomes

30 nm

Section of
chromosome in an
extended form

300 nm

Condensed section
of chromosome

700 nm

Centromere

Entire mitotic
chromosome

1,400 nm

Figure 4.4 Chromatin structure. The base level of compaction is the
11 nm fibre, in which the nucleosomes are positioned as 'beads on a
string'; nucleosomal arrays condense to form the 30 nm fibre of yet unknown structure; higher order compactions lead to the familiar chromosome shape of condensed chromatin. [Reprinted with permission from
Springer Nature from G. FELSENFELD and M. GROUDINE (2003).]

Figure 4.5 Nucleosomal core particle in a ribbon presentation of the core histones bound to DNA (blue). Notice the dangling histone tails. Generated with Chimera from the nucleosome structure PDB 1KX5.

Figure 5.1 Chymotrypsin inhibitor CI2 (PDB entry 1COA). Generated with Chimera. Coiled regions are shown in red, α-helix in green, β-sheet in blue.

Figure 7.1 The λ-phage operon. Operator sites are shown as gray-shaded areas, with OR3-OR2-OR1 from left to right. Promoters are indicated by dashed frames; gene transcription directions are indicated by arrows. Two λ-repressor dimers are shown upon attachment and approach.

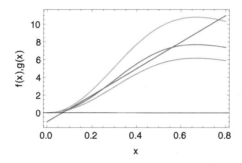

Figure 7.3 Bistability in phage λ. Graph A shows the graphical solution of the stationary points of Eq. (7.12), i.e., $\dot{x} = 0$. In the figure, the three curves are the first term of the rhs of Eq. (7.12), denoted by $f(x)$, while the straight line is given by the function $g(x) = \gamma x - 1$, with $\gamma = 15$. The intersections of the graphs correspond to the extrema of $V(x)$, and can, e.g., be varied by changing γ. In the figure, the green curve is given by $\alpha = 70$, the blue curve by $\alpha = 50$, corresponding to Figure 7.2, while the yellow curve is for $\alpha = 40$.

Figure 7.8 Repressilator oscillations. Shown are the outputs over time.

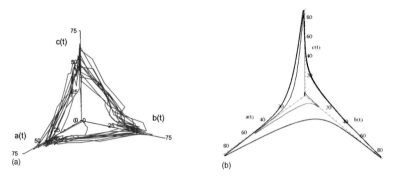

Figure 7.10 (a) The limit cycle of the stochastic repressilator. Simulation parameters are $r = r_p = 1, \varepsilon = 0.1, \eta = 10^{-2}, \delta = 10^{-3}$. (b) The deterministic version for comparison with the reduced system in region III of Figure 7.9, parameters identical to the stochastic version, with Hill coefficient $h = 3$ (R. BLOSSEY and V.G. Giuraniuc, 2008). (Reprinted with permission from the American Physical Society.)

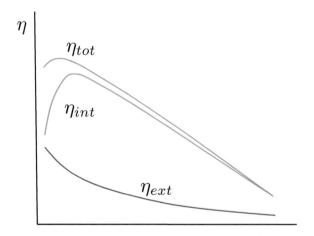

Figure 8.1 *Cellular genetic noise* in *E. coli.* Qualitative behaviour of total, intrinsic and extrinsic noise as a function of fluorescence levels (population means) characterizing transcription rates.

Figure 8.2 MinD, MinE and eGFP-MinC in traveling waves in vitro. (a) Confocal fluorescence micrographs showing waves of MinD, MinE and MinC on a supported lipid membrane. (b) The influence of the presence of MinC on velocity and period of the protein waves. (c) Fluorescence intensity profiles of MinD, MinE and MinC acquired from the rectangular region shown in (a). Starting from the front of the wave (right), the density of MinC rises at a slope similar to that of MinE and also shows a similar sharp decrease at the rear of the wave. The detachment of MinC is shifted toward the front of the wave. [Reprinted with permission from Springer Nature from M. LOOSE et al. (2011).]

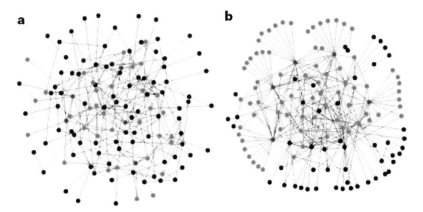

Figure 8.4 Random network, or Erdös-Renyi graph a). The graph under b) represents a scale-free network, discussed in the text below. [Reprinted with permission from Springer Nature (R. ALBERT et al., 2000).]

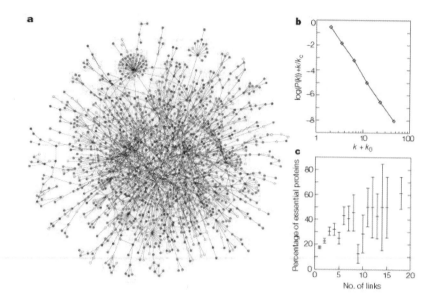

Figure 8.5 A scale-free network of proteins in yeast. a) graphical representation of the network; b) evidence for the fat-tail distribution; c) essentiality of proteins as a function of the number of links. [Reprinted with permission from Springer Nature (H. JEONG et al. 2001).]

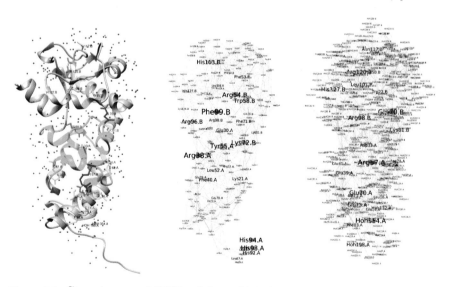

Figure 8.8 Structure and RINs of the colicin E2 DNase-Im2 complex. For all three images, the residues at the interface between the two chains are colored in blue, and central residues are colored in a gradient from yellow (Z-score ≥ 2) to red (Z-score ≥ 4). For the explanation, see the text. (From G. BRYSBAERT et al., 2018.)

II

Non-equilibrium Statistical Mechanics

CHAPTER 5

Back to P: Probabilities over Time

Non-equilibrium statistical mechanics does not yet have such powerful general concepts and tools as equilibrium statistical mechanics. Attempts to develop concepts for physical processes which occur arbitrarily far from a thermal equilibrium situation have so far not led to the desired success.[1]

Since we still lack these general principles which govern the time-dependent or stationary distributions of our physical quantities of interest, we instead have to directly address the properties of the stochastic processes themselves and try to find methods to treat them, at least with some reasonable simplifications.

In the most general case we can address, we will be able to compute the time evolution of suitably defined probability distributions characterizing the state of a physical system, or its transition from one state to another. Consequently, our symbol of choice in this section is P rather than Z.

What we will also see, however, is that we sometimes will be able to relate non-equilibrium processes to equilibrium processes. We will also find expressions, the so-called *fluctuation-dissipation theorems*, which establish such a relation. Most often, however, these relations require for their validity to be not too far away from a thermal equilibrium. There is one notable exception we will address as well, a result obtained some years ago by C. JARZYNSKI. We will deal with these results and their application to biological systems in the subsequent chapter.

---

[1] Of course, there is an extended body of work on *irreversible processes*. However, a theory of non-equilibrium statistical physics of the same level of generality as that of equilibrium statistical physics has not yet been established.

We begin this chapter with the formal bits and pieces of a general description of stochastic processes and their time evolution.[2]

5.1   STOCHASTIC PROCESSES

The first section of this chapter lists a number of definitions for later use.

We call X a *stochastic variable* with a particular value x; f is considered a mapping from X at time t. We call

$$Y_X(t) = f(X, t) \tag{5.1}$$

a *random function*, and $Y_x(t) = f(x, t)$ a *sample function* or a *realization* of the stochastic process; as for equilibrium states, we can speak of the corresponding ensemble in a natural way. For the stochastic process, we define an *ensemble average* by

$$\langle Y(t) \rangle \equiv \int dx \, Y_x(t) \, P(x). \tag{5.2}$$

The higher moments of the distribution are defined in an analogous way. The probability for $Y_x(t)$ to take the value y at time t is given by

$$P(y, t) = \int dx P(x) \delta(y - Y_x(t)). \tag{5.3}$$

The mean and the moments are, as in thermal equilibrium, quantities that allow to quantify the probability distributions. For time-dependent processes, another quantity is of interest, which is the autocorrelation function.

The autocorrelation function. The *autocorrelation function* of the stochastic process is defined by

$$A(t_1, t_2) \equiv \langle Y(t_1) Y(t_2) \rangle - \langle Y(t_1) \rangle \langle Y(t_2) \rangle. \tag{5.4}$$

For $t_1 = t_2 = t$, it reduces to the time-dependent variance, $\sigma^2(t)$.

Joint probability density. We define the *joint probability density*

$$P(y_1, t_1; \ldots; y_n; t_n) \tag{5.5}$$

---

[2]The presentation, in particular at the beginning, follows the book by N. VAN KAMPEN, 1992, which can be seen as a standard reference on most of the topics discussed in this chapter.

which states that $Y_x(t)$ has the value y_1 at t_1,..., y_n at t_n.

Exercise. Write down the mathematical expression for P.

Conditional probability. The *conditional probability* is the probability density for $Y_x(t)$ to take on the value y_2 at t_2 if it was y_1 at t_1, with the normalized density

$$\int dy_2 P(y_2, t_2|y_1, t_1) = 1.$$

(5.6)

Markov process. A stochastic process has the *Markov property* if for any set of ordered timesteps $t_1 < t_2 < ... < t_n$ the conditional probability satisfies

$$P(y_n, t_n|y_1, t_1, ..., y_{n-1}, t_{n-1}) = P(y_n, t_n|y_{n-1}, t_{n-1}).$$

(5.7)

P is then a *transition probability*. It only depends on the two states involved, the one that is left and the one that is reached. A Markov process is uniquely determined from the knowledge of $P(y_1, t_1)$ and $P(y_2, t_2|y_1, t_1)$.

Exercise. Convince yourself of the correctness of the last statement for a process involving three steps, $t_1 < t_2 < t_3$.

The Chapman-Kolmogorov equation. From the transition probabilities of a Markov process we have the expression

$$P(y_3, t_3|y_1, t_1) = \int dy_2 P(y_3, t_3|y_2, t_2) P(y_2, t_2|y_1, t_1)$$

(5.8)

in which time-ordering is essential.

Examples. a) For $-\infty < y < \infty$, the *Chapman-Kolmogorov equation* is solved by (for $t_2 > t_1$)

$$P(y_2, t_2|y_1, t_1) = \frac{1}{\sqrt{2\pi(t_2 - t_1)}} \exp\left[-\frac{(y_2 - y_1)^2}{2(t_2 - t_1)}\right]$$

(5.9)

If $P(y, 0) = \delta(y)$, this Markov process is called the *Wiener* or *Wiener-Lévy process*.

b) If $Y_x(t)$ takes on only positive integer values $n = 0, 1, 2, ...$ for $t \geq 0$, Eq. (5.8) is obeyed by the *Poisson process*

$$P(n_2, t_2|n_1, t_1) = \frac{(t_2 - t_1)^{n_2 - n_1}}{(n_2 - n_1)!} e^{-(t_2 - t_1)}$$

(5.10)

and

$$P(n,0) = \delta_{n,0}\,.\tag{5.11}$$

Task. Show that the probability density of the Wiener process fulfills the *diffusion equation*

$$\partial_t P(x,t) = D\,\partial_x^2 P(x,t)\tag{5.12}$$

with a diffusion constant $D = 1/2$.

Stationary Markov processes. A stochastic process is called *stationary* if all joint probability densities depend only on time differences τ

$$P(y_1,t_1;....;y_n,t_n) = P(y_1,t_1 + \tau; ...; y_n, t_n + \tau)\,.\tag{5.13}$$

In this case we write for the conditional probabilities

$$P(y_2,t_2|y_1,t_1) = T_\tau(y_2|y_1)\tag{5.14}$$

and we will also suppress the index τ whenever no ambiguity can arise. With this notation, the Chapman-Kolmogorov equation is rewritten as

$$T(y_3|y_1) = \int dy_2 T(y_3|y_2) T(y_2|y_1)\,.\tag{5.15}$$

Task. Show that the autocorrelation function of a stationary Markov process with zero mean is given by

$$A(\tau) = \int \int dy_1\, dy_2\, y_1\, y_2\, T_\tau(y_2|y_1)\, P(y_1)\tag{5.16}$$

for $\tau \geq 0$.

Example. The standard example of a stationary Markov process is the *Ornstein-Uhlenbeck process*, for which

$$P(y_1) = \frac{1}{\sqrt{2\pi}} \exp\left(-\frac{y_1^2}{2}\right)\tag{5.17}$$

and

$$T(y_2|y_1) = \frac{1}{\sqrt{2\pi(1 - e^{-2\tau})}} \exp\left[-\frac{(y_2 - y_1 e^{-\tau})^2}{2(1 - e^{-2\tau})}\right]\tag{5.18}$$

where $\tau \equiv t_2 - t_1$.

Exercise. Show that the Ornstein-Uhlenbeck process fulfills $\langle Y \rangle = 0$ and $A(\tau) = e^{-\tau}$.

5.2   THE MASTER EQUATION

We this minimal list of definitions we have set the stage for the description of stochastic processes within the context of statistical physics, and are ready to move on.[3] The next and very important step is to make practical use of the Chapman-Kolmogorov equation for more complex situations than we have discussed so far in the few examples. The corresponding mathematical tool is the *master equation*.

The master equation is a limiting expression one can obtain from the Chapman-Kolmogorov equation if one lets the time interval tend to zero in a controlled way such that one can pass over to a differential equation for the transition probabilities. Let's do this.

Suppose we write down the transition probability T for the case of small time intervals τ' in the following way

$$T_{\tau'}(y_2|y_1) = (1 - a_0(y_1)\tau')\delta(y_2 - y_1) + \tau' W(y_2|y_1) + o(\tau'^2). \qquad (5.19)$$

This is an expansion to linear order in τ', in which $W(y_2|y_1)$ is defined as the transition probability per unit time from y_1 to y_2, hence $W(y_2|y_1) \geq 0$.

The second term in Eq. (5.19) is thus clear, but we have still to say something about the first term. In this term, the coefficient in front of the δ-function takes into account that *no transition occurs* during the time interval interval τ', and hence we must have for the coefficient a_0

$$a_0(y_1) = \int dy_2 W(y_2|y_1). \qquad (5.20)$$

Finally, the last term in Eq. (5.19) denotes terms of higher order in τ' which we neglect in the following.

---

[3]A reader who wishes to get more is asked to consult VAN KAMPEN's book; see the list of References at the end of the chapter.

We insert this expression into the Chapman-Kolmogorov equation and do some rearrangement of terms, with the result

$$\frac{T_{\tau+\tau'}(y_3|y_1) - T_\tau(y_3|y_1)}{\tau'} = -a_0(y_3)\,T_\tau(y_3|y_1) + \int dy_2 W(y_3|y_2)T_\tau(y_2|y_1)$$

(5.21)

which in the limit $\tau' \to 0$ yields the differential equation

$$\partial_\tau T_\tau(y_3|y_1) = \int dy_2 \left[W(y_3|y_2)T_\tau(y_2|y_1) - W(y_2|y_3)T_\tau(y_3|y_1)\right] . \quad (5.22)$$

This is the master equation in its continuous form. Usually it is rewritten in a more intuitive form as

$$\partial_t P(y, t) = \int dy' \left[W(y|y')P(y', t) - W(y'|y)P(y, t)\right] . \quad (5.23)$$

This equation must be solved for $t \geq t_1$ given an initial condition $P(y, t_1) = \delta(t - t_1)$. Note that the equation should not be misinterpreted as an equation for a single-time distribution, which is a frequent error due to the abusive notation with P; the Chapman-Kolmogorov equation makes it clear that we are considering transition probabilities.

Having our tool finally at hand, we can now turn to some illustrative examples. For this, we employ the master equation in a form applicable to discrete states n, and write it as

$$\dot{p}_n(t) = \sum_m \left[w_{nm}p_m(t) - w_{mn}p_n(t)\right] . \quad (5.24)$$

This equation can also be understood as a gain-loss equation for the probability of states n with $w_{nm} \geq 0$.

Example. As an example of a *non-stationary* Markov process, we consider *protein degradation*. Within a cell, proteins are continually degraded by a dedicated cell machinery. In our simple model we describe this by a rate γ. The transition probability is given by

$$w_{nm} = \gamma\, m\, \delta_{n,m-1} \quad (5.25)$$

with $n \neq m$. The master equation of this process reads

$$\dot{p}_n = \gamma(n+1)p_{n+1}(t) - \gamma n p_n(t) \quad (5.26)$$

which needs to be solved under the initial condition $p_n(0) = \delta_{n,n_0}$, i.e., we assume that at time $t = 0$ there are n_0 proteins present.

The simple way to 'solve' Eq. (5.26) is to consider the evolution of the average number of proteins, $\langle n(t) \rangle$. This is done by multiplying the equation by n and summing up

$$
\begin{aligned}
\sum_{n=0}^{\infty} n\dot{p}_n &= \gamma \sum_{n=0}^{\infty} n(n+1)p_{n+1} - \gamma \sum_{n=0}^{\infty} n^2 p_n \\
&= \gamma \sum_{n=0}^{\infty} (n-1)np_n - \gamma \sum_{n=0}^{\infty} n^2 p_n \qquad (5.27) \\
&= -\gamma \sum_{n=0}^{\infty} np_n
\end{aligned}
$$

which is nothing but

$$
\frac{d}{dt} \langle n(t) \rangle = -\gamma \langle n(t) \rangle \qquad (5.28)
$$

solved by

$$
\langle n(t) \rangle = n_0 e^{-\gamma t} \ . \qquad (5.29)
$$

This is, of course, not the full solution to the problem, but it nicely shows how the behaviour of the average can be obtained.

Generating functions. In order to really solve the master equation we can make use of the *generating function*.

$$
G(s,t) = \sum_{n=0}^{n_0} p_n(t) s^n \qquad (5.30)
$$

defined for $|s| \leq 1$. Multiplying Eq. (5.26) by s^n and summing up, the master equation is transformed into a first-order partial differential equation for G in the variables s, t,

$$
\partial_t G(s,t) + \gamma(s-1)\partial_s G(s,t) = 0 \ . \qquad (5.31)
$$

The form of the equation indicates that a *separation of variables* will be of help. We are thus led to write the following *ansatz*[4] for G

$$
G(s,t) = (a(t)(s-1) + b)^{n_0} \ , \qquad (5.32)
$$

---

[4]Note that according to its definition, G is a polynomial in the variable s of order n_0.

where the function $a(t)$ and the constant b need to be determined. Plugging the ansatz into Eq. (5.31), the s-dependence drops out, and the function $a(t)$ is found to fulfill the simple differential equation (compare to Eq. (5.28))

$$\frac{\dot{a}(t)}{a(t)} = -\gamma \tag{5.33}$$

with the solution $a(t) = e^{-\gamma t}$. The constant $b = 1$, as follows from the initial condition $G(s,0) = s^{n_0}$. Thus

$$G(s,t) = (1 + e^{-\gamma t}(s-1))^{n_0} . \tag{5.34}$$

Now we have two expressions for $G(s,t)$, the power series in s and its sum. Since the $p_n(t)$ are the coefficients of the series, we only have to Taylor expand Eq. (5.34), which leads to the final result

$$p_n(t) = \binom{n_0}{n} \exp(-\gamma n t)(1 - \exp(-\gamma t))^{n_0-n} . \tag{5.35}$$

One-step processes. A frequently occurring class of stochastic processes that can be studied with the help of master equations are *one-step processes* in which transitions occur only between neighbouring state labels. Writing the master equation as a matrix equation[5]

$$\dot{p}_n = \mathcal{W}_{nm} p_m \tag{5.36}$$

we define the matrix $\mathcal{W}_{nm}$ as

$$\mathcal{W}_{nm} = f_m \delta_{n,m-1} + b_m \delta_{n,m+1} \tag{5.37}$$

with $n \neq m$. The diagonal element of the matrix is given by

$$\mathcal{W}_{nn} = -(f_n + b_n) \tag{5.38}$$

so that the full equation reads as

$$\dot{p}_n = f_{n+1} p_{n+1} + b_{n-1} p_{n-1} - (f_n + b_n) p_n . \tag{5.39}$$

---

[5]Which makes the linearity of the equation evident.

The last equation can be written in a more concise form using the *step operator* $\mathcal{E}$, defined by its action to "move up or down the ladder" of states given by h_n:

$$\mathcal{E}h_n = h_{n+1} , \quad \mathcal{E}^{-1}h_n = h_{n-1} \tag{5.40}$$

so that for Eq. (5.39) we have

$$\dot{p}_n = (\mathcal{E} - 1)f_n p_n + (\mathcal{E}^{-1} - 1)b_n p_n . \tag{5.41}$$

Exercise. Solve the master equation (5.39) for the *symmetric random walk* with $f_n = b_n = 1$,

$$\dot{p}_n = p_{n+1} - p_{n-1} - 2p_n , \quad -\infty < n < \infty \tag{5.42}$$

for the initial data $p_n(0) = \delta_{n,0}$.

Folding of small proteins. As a final example for the master equation we discuss small proteins that are *two-state folders*. Ideally, a two-state folder should have exactly two states: a denatured (coil) state, and the native folded (helix) state. If this is the case, the discussion of the denaturation of short DNA chains applies: native and denatured state are describable as a chemical equilibrium; the only difference being the calculation of the free energy. So we are done.

If true, this would imply that the folding of small proteins is about as feature-less as the denaturation/hybridization of a short DNA fragment, and only the melting temperature (here, the folding temperature) would depend somewhat on amino acid composition. We would then expect the following under *muta-tions* of the amino acid composition of the chain: suppose we replace amino acids and look at the folding profile in the same way as we did for the melt-ing profile of DNA. If the amino acids have only different binding strengths, the resulting curves should be shifted in temperature but otherwise not much affected.

But that is not what is observed. Studies of protein unfolding kinetics have uncovered quite a different picture. The effect of mutations on folding can be quantified by the so-called Φ-*value*

$$\Phi \equiv \frac{RT \ln(k_{wt}/k_m)}{\Delta G_M} , \tag{5.43}$$

where k_{wt} and k_m are the folding rates of the wild-type and mutated proteins, and ΔG_M is the free energy associated with the change in thermal stability due to the mutation. Φ is also related to the free energy difference between the *transition state* and the native state in the non-mutated protein, ΔG_T. One writes $\Phi = \Delta G_T/\Delta G_M$, neglecting prefactor differences.

Φ is found to have values between 0 and 1, indicating that the residues in the ensemble of *transition states*. These are those pseudo equilibrium structures between the folded or unfolded equilibrium structures which have at least a partial native-like structure. Sometimes values less than 0 or larger than 1 appear which are not so easy to interpret. In fact, the most important observed feature of Φ is that changes in neighbouring bases can have very different effects on the Φ-value. Φ thus does not vary in a continuous manner upon mutations along the chain.

In hindsight this is not surprising, given that the various amino acids have rather different chemical nature, in contrast to the nucleic acids.

Let's discuss this issue in more detail, for one particular example, the chymotrypsin inhibitor CI2. Its structure is shown in Figure 5.1. We base the discussion on a simple model specifically tuned to this case. The molecule CI2 has one α-helix and a four-stranded β-sheet. In the α-helix, 20 single residue mutations have been studied experimentally with Φ-values in the range between $-0.35 \leq \Phi \leq 1.25$. What does a simple model look like to reproduce these data, at least the underlying trend?

In a very simple *ansatz* one can attempt to describe the folding kinetics by a master equation

$$\dot{\mathbf{p}}(t) = \mathbf{W}\mathbf{p}(t), \tag{5.44}$$

where $\mathbf{p}$ is the vector of the state probabilities of the protein. We assume the transition state from a state m to a state n to be given by

$$w_{nm} = \frac{1}{t_0} \left[\frac{1}{1 + exp[(G_n - G_m)/RT]} \right], \tag{5.45}$$

where G_i is the free energy of each partially folded state, and t_0 is a characteristic time.

The solution of Eq. (5.44) is given by

$$\mathbf{p}(t) = \sum_\lambda \mathbf{Y}_\lambda \exp[-\lambda(t/t_0)] \tag{5.46}$$

with eigenvectors Y_λ and eigenvalues λ. C. MERLO et al., 2005 assumed the four folding states:

- the denatured state (D);

- a partially folded state (α) with a α-helix;

Figure 5.1 (**See color insert.**) Chymotrypsin inhibitor CI2 (PDB entry 1COA). Generated with Chimera. Coiled regions are shown in red, α-helix in green, β-sheet in blue.

- a partially folded state (β) with a β-sheet;
- the native state (N).

Hence in this case, the matrix $\mathbf{W}$ is a (4×4)-matrix.

Due to its simplicity, the model can be solved exactly, and gives the eigenvalues

$$\lambda_{0,...,3} = (0, \ 1 - q, \ 1 + q, \ 2) \tag{5.47}$$

with

$$q = \frac{1 - \alpha\beta/N}{\sqrt{(1 + \alpha)(1 + \beta)(1 + \alpha/N)(1 + \beta/N)}} \tag{5.48}$$

with $-1 < q < 1$, and where $\alpha \equiv e^{-G_\alpha/RT}$; analogously for D, N and β.

Despite its four states, the model is found to display two-state kinetics provided the free energy of the native state is much smaller than the free energy of the other states and, further, if the α- and β-intermediaries have free energies larger than D. In this case a kinetic barrier between D and N exists, and the transition is governed by a single-exponential dynamics.

Given these conditions, q in Eq. (5.48) can be simplified to (*Exercise*)

$$q \approx ((1+\alpha)(1+\beta))^{-1/2} , \tag{5.49}$$

and the folding rate k - i.e., the smallest relaxation rate λ_0 - simplifies upon expansion of the square-root for $\alpha, \beta \ll 1$ to

$$k \equiv 1 - q \approx \frac{\alpha + \beta}{2} . \tag{5.50}$$

From this result one infers that the folding rate is simply the sum of the rates of folding into either the α- and or β-substructures with equal probability.

The model can now be used to study the effect of mutations. Treating the effect of the mutations on the free energies as small perturbations ΔG one can write

$$\ln \left(\frac{k_{wt}}{k_m} \right) \approx \frac{\partial \ln k}{\partial \ln G_\alpha} \Delta G_\alpha \tag{5.51}$$

with k taken from Eq. (5.50). Consequently, with $i = \alpha, \beta$,

$$\Phi = \chi_i \frac{\Delta G_i}{\Delta G_N} \tag{5.52}$$

with, e.g., for $i = \alpha$

$$\chi_\alpha = -RT \frac{\partial \ln k}{\partial G_\alpha} = \frac{\alpha}{\alpha + \beta} . \tag{5.53}$$

Within this simple model one finds that Φ is the product of a *structural factor* χ and an *energetic factor* $\Delta G_i / \Delta G_N$. Although χ obviously involves energies, it can nevertheless be considered structural since it explicitly depends on the possible intermediaries, be they the formation of an α-helix or a β-sheet.

Coming back to the protein CI2, the experimental Φ-values of the twenty mutations on the α-helix could indeed be reproduced with this model with a satisfactory correlation coefficient of 0.85 (C. MERLO et al., 2005). It should be kept in mind, however, that the calculation is indeed a simplified one: in reality there may well be interactions (so-called *tertiary contacts*) between the α-helix and the β-sheet; the model assumes that both fold independently from each other.

5.3   THE FOKKER-PLANCK AND LANGEVIN EQUATIONS

In this section, we discuss two further classic analytical approaches commonly used to describe the stochastic dynamics of particle systems.

The Fokker-Planck equation. In this subsection we will introduce a continuous approximation to the master equation. First, we recall the last general expression we had obtained for it, Eq. (5.24)

$$\partial_t P(y,t) = \int dy' \left[W(y|y')P(y',t) - W(y'|y)P(y,t) \right] . \tag{5.54}$$

We now rewrite the transition probabilities W as

$$W(y|y') = W(y';r) , \qquad r = y - y' \tag{5.55}$$

and obtain

$$\partial_t P(y,t) = \int dr W(y-r;r)P(y-r,t) - P(y,t) \int dr W(y;-r) . \tag{5.56}$$

This is just a rewrite, but now we want to play with the difference variable $r = y - y'$ which will allow us to formulate a continuum approximation.

In order to formulate this continuum version of the master equation we assume for the dependence of the transition probabilities on r that they fulfill the following

$$W(y';r) \approx 0, \quad |r| > \delta \tag{5.57}$$
$$W(y' + \Delta y; r) \approx W(y';r), \quad |\Delta y| < \delta \tag{5.58}$$

which means that W is a slowly varying function of y', but sharply peaked in its dependence on r. If, additionally, P varies also slowly with y, we can expand in a Taylor series to obtain

$$\partial_t P(y,t) = \int dr W(y;r)P(y;t) - \int dr \, r \, \partial_y [W(y;r)P(y,t)]$$

$$\tag{5.59}$$

$$+ \frac{1}{2} \int dr \, r^2 \partial_y^2 [W(y;r)P(y,t)] - P(y,t) \int dr W(y;-r)$$

where the first and the last term on the right-hand side (rhs) cancel each other out. The remaining integrals over r can be absorbed in the definition of the *jump moments*

$$a_\alpha = \int_{-\infty}^{\infty} dr\, r^\alpha W(y; r) \qquad (5.60)$$

and we finally obtain

$$\partial_t P(y, t) = -\partial_y [a_1(y) P(y, t)] + \frac{1}{2} \partial_y^2 [a_2(y) P(y, t)]. \qquad (5.61)$$

This is the *Fokker-Planck equation*. It contains a *drift term* $\sim \partial_y P$ and a *diffusion term*, $\sim \partial_y^2 P$. If we suppose a stationary distribution, $\partial_t P = 0$, drift and diffusion term have to balance each other.

Example for the Fokker-Planck equation: Brownian motion. Consider a particle suspended in a liquid. If we trace its motion under the influence of random molecular collisions of the liquid, we find a continuous path, as shown in Figure 5.2. Measuring the distances the particle travels and averaging over several realizations we obtain for the jump moments the expressions

$$a_1 \equiv \frac{\langle \Delta x \rangle_X}{\Delta t} = 0, \quad a_2 \equiv \frac{\langle (\Delta x)^2 \rangle_X}{\Delta t} = const. \qquad (5.62)$$

Thus, the Brownian particle obeys a diffusion equation

$$\partial_t P(x, t) = \frac{a_2}{2} \partial_x^2 P(x, t) \qquad (5.63)$$

and we can identify $a_2/2 \equiv D$, i.e. the diffusion coefficient is given by

$$D = \frac{\langle (\Delta x)^2 \rangle}{\Delta t}. \qquad (5.64)$$

We now want take a second, different look at the Brownian particle.[6] Let us now consider the velocity instead of the position of a suspended particle as the dynamic variable. We first ignore the random collisions of the particle of the solution molecules and assume that the velocity of the particle relaxes according to

$$\dot{v} = -\gamma v, \qquad (5.65)$$

---

[6]What we consider here is actually called the *Rayleigh particle* which is equivalent to the Brownian particle; the difference between the two is the fine graining of the time scale. In the discussion of the Rayleigh particle one assumes that $t_{coll} \ll \Delta t \ll t_{relax}$, where t_{coll} is a molecular collision time, and t_{relax} the time scale on which the particle velocity relaxes.

Figure 5.2 Particle trajectory in solution: a Brownian path.

i.e., the particle velocity will go to zero for $t \to \infty$: the differential equation is the same as that for the mean number of proteins in the protein degradation problem we discussed with help of the master equation.

Hence, there is a now a drift term in the corresponding Fokker-Planck equation, and it is given by the jump moment

$$a_1(v) = \frac{\langle \Delta v \rangle_V}{\Delta t} = -\gamma v . \tag{5.66}$$

The second jump moment fulfills $a_2(v) > 0$ even for $v = 0$. The Fokker-Planck equation reads as

$$\partial_t P(v,t) = \gamma \partial_v [v P(v,t)] + \frac{a_2}{2} \partial_v^2 P(v,t) . \tag{5.67}$$

The stationary distribution for this process is actually known from equilibrium statistical mechanics: it is the *Maxwell-Boltzmann velocity distribution* in which m is the particle mass and $\beta = 1/k_B T$,

$$P(v) = \left(\frac{\beta m}{2\pi}\right)^{1/2} \exp\left(-\frac{\beta m v^2}{2}\right) . \tag{5.68}$$

With the help of this expression, we can identify the coefficient of the diffusion term as $a_2/2 = \gamma/(m\beta)$, and the Fokker-Planck equation is fully given by

$$\partial_t P(v,t) = \gamma \partial_v \left[v P(v,t) + \frac{1}{\beta m} \partial_v P(v,t) \right] . \tag{5.69}$$

Exercise. Compute $\langle v(t) \rangle$ and $\langle v^2(t) \rangle$ from Eq. (5.69) for the given initial velocity $v(0) = v_0$.

Exercise. Compute the dissipation-fluctuation relation for the Brownian motion from the Fokker-Planck equation for $P(v,t)$.

The Langevin equation. We now complete the discussion of the Brownian motion of the suspended particle in the velocity description by the classic approach originally suggested by P. LANGEVIN.

In order to do this we consider the velocity process under the inclusion of a noise source $\eta(t)$ which we add to the rhs of Eq. (5.65),

$$\dot{v} = -\gamma v + \eta(t).\tag{5.70}$$

The noise source η models the random action of the solution molecules on the suspended particle (we now put $m = 1$). The solution particles 'kick' the suspended particle and transfer part of their thermal energy. For the first two moments of the noise distribution we assume

$$\langle \eta(t) \rangle = 0, \quad \langle \eta(t)\eta(t') \rangle = \Gamma\delta(t - t'),\tag{5.71}$$

where Γ is a constant. The motivating idea behind this specific choice is that the random collisions are instantaneous and uncorrelated.

Exercise. How would the rhs of the second moment change if one were to introduce a small but finite collision time τ_c?

If we assume an initial velocity $v(0) = v_0$ of the particle, as we did before, we can compute the velocity $v(t)$ as

$$v(t) = v_0 e^{-\gamma t} + e^{-\gamma t}\int_0^t dt e^{\gamma t'}\eta(t')\tag{5.72}$$

and, using the moments of η, we can obtain those of v as:

$$\langle v \rangle = v_0 e^{-\gamma t}\tag{5.73}$$

$$\langle v^2(t) \rangle = v_0^2 e^{-2\gamma t} + \frac{\Gamma}{2\gamma}(1 - e^{-2\gamma t}).\tag{5.74}$$

We are now left to determine the coefficient Γ. For $t \to \infty$, we know from Eq. (5.68) - for $m = 1$ - that

$$\langle v^2(t \to \infty) \rangle = k_B T = \frac{\Gamma}{2\gamma} . \tag{5.75}$$

Again we have found an example of the *dissipation-fluctuation theorem*. A further, very prominent example is the *Einstein relation*

$$D = \frac{k_B T}{\gamma} , \tag{5.76}$$

see the *Exercise* on the Fokker-Planck equation in velocity space.

Fokker-Planck vs. Langevin. The Fokker-Planck equation determines the full stochastic process of the Brownian particle; by contrast, the Langevin equation does, by construction, not go beyond second moments. Therefore, if we additionally assume that the noise η is Gaussian, all odd moments will vanish, and the even moments will factorize (*Exercise!*). This leads to what is commonly called a *Gaussian white noise*, i.e., a noise spectrum containing all frequencies, which usually serves to model a rapidly fluctuating force.

This construction is all fine for the simple case of a Brownian particle, and lends itself to the many applications of the Langevin equation approach, in which a Gaussian white noise is added to the known deterministic equations of the system. The constant Γ is then usually adjusted such that the stationary solution correctly matches with the fluctuations around the stationary state.

But it is worth keeping in mind that this approach is a strong simplification, as has been advocated by N. VAN KAMPEN. We will briefly go through the main points here; for a more detailed discussion the reader is referred to van Kampen's book. Furthermore, we will return to this discussion in Part II when we will discuss the role of fluctuations in biological systems in Chapter 6.

What has to be kept in mind if one wants to treat stochastic fluctuations with the Langevin approach? By this we refer to the procedure we used for the description of Brownian motion, namely to first write down a deterministic equation for the macroscopic (average) dynamics, and to then add on the fluctuations.

i) Suppose your system is described by a deterministic equation of the type $\dot{u} = A(u) = \sin u$ to which we add the noise as defined above,

$$\partial_t u = \sin u + \eta(t) . \tag{5.77}$$

If we average this equation, we find

$$\partial_t \langle u \rangle + \langle \sin u \rangle = 0 \tag{5.78}$$

since, as before, we have $\langle \eta(t) \rangle = 0$. This result, however, means that the average does not obey the macroscopic equation, i.e., $\partial_t \langle u \rangle \neq \sin \langle u \rangle$, since

$$\langle \sin u \rangle = \sin \langle u \rangle - \frac{1}{2} \langle (u - \langle u \rangle)^2 \rangle \cos \langle u \rangle + ... , \tag{5.79}$$

which is an equation involving all higher moments.

The message of this calculation obviously is: if we start from a deterministic equation for the average which is *nonlinear*, the simple addition of fluctuating source will in general be too naive.

ii) For an arbitrary nonlinearity $A(u)$ the Langevin equation Eq. (5.77) is equivalent to the Fokker-Planck equation

$$\partial_t P(u,t) = -\partial_u (A(u) P(u,t)) + \frac{\Gamma}{2} \partial_u^2 P(u,t) . \tag{5.80}$$

If we allow equations of the type

$$\partial_t u = A(u) + C(u) \eta(t) \tag{5.81}$$

in which a u-dependent function multiplies the noise, we run into an interpretation problem. For each jump in the solution u of the equation due to the noise, the value of u and hence of $C(u)$ is undetermined. We thus have to specify a rule how to interpret the product $C(u)\eta(t)$, and this leads to a dependence of the resulting Fokker-Planck equation on that rule. Some possibilities are to take the value of $C(u)$ before the jump, after the jump or the mean. The different options lead to different Fokker-Planck equations and hence to different results.[7] The option to take the mean value is named after R. L. STRATONOVICH, the version to take the value before the jump is named after K. ITÔ.

We illustrate this phenomenon for the example of protein degradation for which we had the master equation

$$\dot{p}_n = \gamma(n+1) p_{n+1}(t) - \gamma n p_n(t) . \tag{5.82}$$

---

[7]The two stochastic equations differ by what is called a *spurious drift term*, and the equations can be transformed into each other by corresponding transformation rules; see the detailed discussion in van Kampen's book.

Within the Langevin approach we can assume

$$\dot{n} = -\gamma n + \sqrt{n}\eta(t) \tag{5.83}$$

since we expect the fluctuations in the decay process to be proportional to the square root of the number of proteins. Thus, we have a case in which we need to determine the multiplication rule for the noise.

In the Itô-case, we take the value before the jump, hence

$$n(t + \Delta t) - n(t) = -\gamma n(t)\Delta t + \sqrt{n} \int_t^{t+\Delta t} dt' \eta(t'). \tag{5.84}$$

It can be proved that this choice leads to the Fokker-Planck equation

$$\partial_t P(n,t) = \partial_n (\gamma + \frac{\Gamma}{2}\partial_n)[nP(n,t)]. \tag{5.85}$$

The corresponding line of reasoning in the Stratonovich case yields

$$\partial_t P(n,t) = \partial_n \left[\left(\gamma n - \frac{\Gamma}{2}\right) P + \frac{\Gamma}{2}\partial_n [nP(n,t)] \right]. \tag{5.86}$$

If we calculate the equation for the average $\langle \dot{n} \rangle$ from the Itô-version of the Fokker-Planck equation, we obtain the same result as Eq. (5.28) which we obtained directly from the master equation (*Exercise*). The Stratonovich equation, by contrast, does not yield this result. The reason is that in the process of protein degradation, the probability to go from n to $n-1$ proteins is indeed proportional to the number n of available proteins before the transition, in line with the Itô assumption of the construction of the stochastic process in the first place.

The ambiguity arising from the different possible choices is referred to as the *Itô-Stratonovich dilemma*. A better understanding of what is an adequate procedure in formulating stochastic equations can be obtained from a distinction between *intrinsic* and *extrinsic* noise sources. Extrinsic noise is an 'add-on' to a fundamentally deterministic dynamics, as it is the case in many engineering applications. Here, one models, e.g., an electrical circuit on the level of its macroscopic constituents like resistors and capacitors and not on the level of electrons. In such a case, noise is never really white but has a finite correlation time. In this case, there is no ambiguity and the Stratonovich prescription applies.

Intrinsic noise, by contrast, is due to the fact that the system itself is composed of discrete particles which interact stochastically: this noise can *never* be switched off, and hence $A(u)$ is not determined by the evolution equations in a system isolated from noise. Hence, the Langevin approach does in general not work for internal noise. Instead, one has to go back to the master equation approach, from which in certain cases macroscopic equations can be derived. This derivation can be done as follows in a systematic way. The starting point is the probability density $P(y, t)$ for a stochastic process Y_X. We define

$$y(t) \equiv \langle Y \rangle_t = \int dy \, y \, P(y, t) \tag{5.87}$$

where we assume that the density P is a sharply peaked function of its argument y; it is assumed to be initially of the form $P(y, 0) = \delta(y - y_0)$.

As t increases, the probability density P will evolve along y, and this determines the behaviour of the average $y(t)$. One finds

$$
\begin{aligned}
\dot{y}(t) &= \int dy \, y \, \partial_t P(y, t) \\
&= \int dy \, dy' \, y \left[W(y|y') P(y', t) - W(y'|y) P(y|t) \right] \tag{5.88} \\
&= \int dy \, dy' (y' - y) W(y'|y) P(y, t). \tag{5.89}
\end{aligned}
$$

We now recall the *jump moments* (Eq. (5.60))

$$a_\alpha(y) = \int dy' (y' - y)^\alpha W(y'|y) \tag{5.90}$$

for $\alpha = 0, 1, 2, ..$, to obtain

$$\dot{y}(t) \equiv \frac{d}{dt} \langle Y \rangle_t = \int dy \, a_1(y) P(y, t) = \langle a_1(Y) \rangle_t \tag{5.91}$$

If we expand $a_1(y)$ around the average $\langle Y \rangle_t$ in a Taylor series

$$\langle a_1(Y) \rangle_t = a_1(\langle Y \rangle_t) + \frac{1}{2} \langle (Y - \langle Y \rangle_t)^2 \rangle_t a_1''(\langle Y \rangle_t) + ... \tag{5.92}$$

one finds to a first approximation the equation

$$\dot{y}(t) = a_1(y(t)) = A(y) \tag{5.93}$$

which is, despite of P fulfilling a linear equation, in general a *nonlinear* equation of $y(t)$, even when the higher-order terms are dropped.

Task. Compute the corrections to the macroscopic equations. For this, consider

$$\frac{d}{dt}\langle Y^2(t)\rangle_t \tag{5.94}$$

and the variance $\sigma^2(t) = \langle Y^2(t)\rangle - y^2(t)$. For jump moments a_1 and a_2 which are linear in y show that the macroscopic equation can be written as

$$\dot{y}(t) = a_1(y(t)) + \frac{1}{2}\sigma^2(t)a_1''(y(t)). \tag{5.95}$$

A generalization of Z. As a final step in our development of the theory of stochastic processes we discuss a recent observation by (T. B. LIVERPOOL, 2018), who showed that for steady-state stochastic processes a generalized distribution can be derived for the average trajectories. Starting point of the discussion is the Langevin equation

$$\dot{X}_i = -D_{ij} \cdot (\nabla \mathcal{H}(X_i) + v_i(X_i)) + \xi_i \tag{5.96}$$

where X_i is a vector with components $i = 1, .., N$, D a mobility matrix and $v_i(X_i)$ a velocity term which cannot be derived from a scalar function, hence explicitly breaks detailed balance. The noise ξ_i is Gaussian white noise with a variance

$$\langle \xi_i(t)\xi_j(t')\rangle = 2\theta D_{ij}\delta(t - t'). \tag{5.97}$$

We consider $D_{ij} = D_i\delta_{ij}$ with $D_i > 0$ independent of X_i. The parameter $\theta > 0$ will play the role of thermal energy, k_BT, in the equilibrium case, as we will see below.

The Langevin equation is equivalent to a Fokker-Planck equation given by

$$\partial_t P = \sum_{i=1}^{N} \nabla_i D_i(\theta\nabla_i P + P(\nabla_i \mathcal{H} - v_i)). \tag{5.98}$$

According to (T. B. LIVERPOOL, 2018), this equation has a stationary distribution given by

$$P_{ss} = \varrho(\mathbf{X}) = \frac{1}{Z}e^{-h(\mathbf{X})} \tag{5.99}$$

with $Z = \int d^N X e^{-h}$, in perfect analogy to the equilibrium case, provided that the condition

$$\sum_{i=1}^{N} D_i L_i(h) = 0 \tag{5.100}$$

is fulfilled for the function

$$L_i(h) = \theta(\nabla_i h)^2 + \nabla_i^2 \mathcal{H} + \nabla_i h(v_i - \nabla_i \mathcal{H}) - \theta(\nabla_i^2 h) - \nabla_i v_i. \qquad (5.101)$$

The stationary density ϱ is then constant on the trajectories

$$\dot{X}_i = V_i = D_i(v_i - \nabla_i \mathcal{H} + \theta \nabla_i h). \qquad (5.102)$$

In the equilibrium case we have for the velocity $v_i = 0$ as well as $V_i = 0$, and hence find

$$h = \frac{\mathcal{H}}{\theta} \qquad (5.103)$$

which justifies the identification of $\theta = k_B T$ in equilibrium. When $\theta = 0$, the average trajectories are those of the deterministic equation. As other trajectories do not keep the probability density constant, these trajectories act as attractors. The condition on the eigenvalues of the Hessian matrix for h is that $\nabla_i \nabla_j h > 0$, i.e. their strict positivity.

Exercise. Illustrate this idea for a simple chemical oscillator, the *Brusselator*, an oscillatory chemical reaction with two components. Their dynamics reads as

$$\dot{x} = \mu + x^2 y - \lambda x - x + \xi_1(t) \qquad (5.104)$$

and

$$\dot{y} = \lambda x - x^2 y + \xi_2(t). \qquad (5.105)$$

Compute h. *Hint*: write $h(x, y)$ as a power series in x, y.

This concludes our discussion of the general methods of stochastic dynamics, i.e., master equations, Fokker-Planck and Langevin equations. We now move on to a class of systems to which these approaches will be applied later, to biochemical reaction systems, and give the basic elements of the theory of chemical kinetics. We discuss two approaches: first, the deterministic one, based on rate equations. Subsequently, we return to the stochastic aspect by a description of the *Gillespie algorithm* which allows to simulate the master equation for biochemical reactions.

Chemical kinetics. In many applications of stochastic processes in biology, we have to deal with biochemical reactions. Consider thus a system of m chemical substances, and n *chemical reactions*. We represent the latter by n *collision diagrams*, given by

$$\nu_{i1} A_1 + \dots + \nu_{im} A_m \xrightarrow{k_i} \mu_{i1} A_1 + \dots + \mu_{im} A_m \qquad (5.106)$$

for $i = 1, .., n$. A_i can represent either the number of a molecule of substance i, or its concentration (obtained by division through volume, $n_i = A_i/\Omega$). The k_i are *rate constants*; ν_{im}, μ_{im} are stoichiometric coefficients, which are integer numbers. The sums

$$r_i = \sum_{j=1}^{m} \nu_{ij} \tag{5.107}$$

define the *order of the transitions*. A transition

$$A_1 + A_2 \to^k A_3 \tag{5.108}$$

thus represents the binary collision of A_1 and A_2, giving rise to A_3. Higher order collisions have a low probability to occur, and are generally neglected in chemical kinetics.

In order to pass from a *collision diagram* to equations describing the time evolution, several assumptions have to be made. We begin with the following:

- the system (typically, a solution) can be assumed spatially homogeneous;

- the density of the molecules is low;[8]

- all reactions occur at a constant volume and temperature;

- the collision frequency of the molecules depends on local concentration.

We can then write down the equation for the chemical kinetics of the molecules in the form

$$\frac{dA_j}{dt} = \sum_{i=1}^{n} k_i(\mu_{ij} - \nu_{ij})A_1^{\nu_{i1}} \cdots A_m^{\nu_{im}} \tag{5.109}$$

for $j = 1, ..., m$. This expression is the *rate equation*.

The equations expressed in (5.109) are not always independent. This can be made more explicit by considering them in matrix form

$$\frac{d}{dt}\mathbf{A} = \mathbf{M} \cdot \mathbf{K} \tag{5.110}$$

where A_j is a vector with components A_j, $\mathbf{M}$ is a matrix with row vectors $\mathbf{V}^T \equiv (\mu_{1j} - \nu_{1j}, ..., \mu_{nj} - \nu_{nj})$ for $j = 1, ..., m$, and $\mathbf{K}^T \equiv (k_1 A_1^{\nu_{i1}} \cdot \cdots A_m^{\nu_{1m}}, ..., k_n A_n^{\nu_{n1}} \cdots A_m^{\nu_{nm}})$. If the matrix $\mathbf{M}$ has a rank r with $r \leq \min\{n, m\}$

---

[8]But not too low; if the molecules appear in too few numbers, the continuum approach breaks down.

the equations are linearly dependent; there are thus $m - r$ *conservation laws* which can be written as

$$\sum_{j=1}^{m} \alpha_{jk} A_j(t) = \sum_{j=1}^{m} \alpha_{jk} A_j(0) \tag{5.111}$$

for $k = 1, ..., m - r$.

Task. Consider Eq. (5.109) for the case of one collision diagram, with reaction and back-reaction rate constants k_+ and k_-.

a) Write down the *chemical master equation* using the step operator $\mathcal{E}$.

b) For an ideal mixture of molecules, the grand canonical distribution is given by

$$P(\{n_i\}) = \prod_{j} \frac{(\Omega z_j)^{n_j}}{n_j!} \exp(-\Omega z_j) \tag{5.112}$$

with $n_j = 0, 1, ...$ Here

$$z_j = \left(\frac{2\pi m}{\beta}\right)^{3/2} \sum_{\nu} \exp(-\beta \epsilon_\nu) \tag{5.113}$$

is the partition function of one molecule j in a unit volume. The energies ϵ_ν contain all internal molecular degrees of freedom, be they rotational, vibrational or electronic. Check that $P(\{n_i\})$ is a stationary solution of the master equation obtained under a), if

$$\frac{k_+}{k_-} = \prod_{j} z_j^{\mu_j - \nu_j} \tag{5.114}$$

applies, i.e., the *law of mass action* is fulfilled.

The Gillespie algorithm. Since master equations are usually easy to write down but hard to solve explicitly, D. T. GILLESPIE, 1977 proposed a simple prescription how to determine the time evolution corresponding to the master equation from a stochastic algorithm.

Starting point of the algorithm is the expression for the transition probability of the Chapman-Kolmogorov equation, Eq. (5.21), in the case of small time interval τ. The algorithm can be formulated in terms of the probability P for a given reaction $i = 1, ..., n$ to occur in an infinitesimal time-interval $d\tau$, which can be written as

$$P(\tau, i) = P_0(\tau) \cdot a_i \cdot d\tau \tag{5.115}$$

where $P_0(\tau)$ is the probability that *no* collision will have occurred in the interval $(t, t + \tau)$, while $a_i \equiv h_i c_i$ is a *stochastic reaction rate* c_i, multiplied by a combinatorial factor determined by the type of reaction that occurs, and by counting the number of distinct reaction partners that are available for a reaction. Thus, $a_i \cdot d\tau$ is the probability that a collision will occur in the interval $(t + \tau, t + \tau + d\tau)$. Defining

$$a_0 \equiv \sum_{i=1}^{M} a_i, \qquad (5.116)$$

$P_0(\tau)$ obeys the equation

$$P_0(\tau + d\tau) = P_0(\tau) \cdot [1 - a_0 \cdot d\tau] \qquad (5.117)$$

which describes the (obvious) fact that the probability that *no* reaction occurred in time $\tau + d\tau$ equals the product of the probabilities that no transition occurred in the interval τ, and within $d\tau$.

The solution of this equation is given by $P_0(\tau) = \exp(-a_0\tau)$, such that

$$P(\tau; i) = a_i \exp(-a_0\tau), \quad 0 \leq \tau < \infty, i = 1, .., n, \qquad (5.118)$$

or 0 otherwise.

This result can be used to compute the time evolution of the system by the reconstruction of $P(\tau; i)$ from a draw of two random numbers r_1 and r_2 from the unit interval uniform distribution according to the prescription

$$\tau = \frac{1}{a_0} \ln \left(\frac{1}{r_1} \right), \quad i = r_2 \qquad (5.119)$$

where i is selected such that the condition

$$\sum_{j=1}^{i-1} a_j < r_2 a_0 \leq \sum_{j=1}^{i} a_j \qquad (5.120)$$

is fulfilled.

These steps can finally be cast into the following algorithm:

- define the stochastic rates c_i for $i = 1, ..., n$, and the initial conditions on the N molecules;

- calculate $a_1 = h_1 c_1, ..., a_n = h_n c_n$, and $a_0 = \sum_{j=1}^{n} a_j$;

- generate r_1, r_2 and calculate τ and i;

- increase t by τ, adjust the population levels of the selected reaction i; loop.

5.4 SEQUENCE ALIGNMENT: A NON-EQUILIBRIUM PHASE TRANSITION

We will now discuss a phase transition, but not an equilibrium one as we did in Chapter 1. We will find critical exponents and scaling, and we will both use stochastic equations in their discrete and continuous versions. The topic is *sequence alignment*, a fundamental method in computational biology developed to find out the similarities between two sequences, be they made up DNA bases, or the amino acid sequences of proteins.

In this section, we will see that the alignment of two sequences can be understood as a phase transition in a non-equilibrium system. This may, at first sight, seem astonishing: what does a pattern matching problem have to do at all with a phase transition? There are several aspects that have to be addressed in order to answer this question, and we will do so as we go along, and follow the exposition by R. BUNDSCHUH, 2002.

We begin with a technical definition of the notion of an alignment.

Gapless alignment. The simplest procedure of sequence alignment is called *gapless alignment*. It looks for similarities between two sequences $\mathbf{a} = (a_1, a_2, ..., a_M)$ and $\mathbf{b} = (b_1, b_2, ..., b_N)$ where $M \sim N$. The letters a_i, b_i are taken from an alphabet of size c; for our purposes here we take $c = 4$, i.e., the four-letter alphabet of the DNA bases.

A local gapless alignment A consists of two substrings of equal length ℓ of the sequences $\mathbf{a}, \mathbf{b}$. To each such alignment can be assigned a *score*

$$S[A] = S(i, j, \ell) = \sum_{k=0}^{\ell-1} s_{a_{i-k}, b_{j-k}} \tag{5.121}$$

where the *scoring matrix* is given, in the simplest case, by

$$s_{a,b} = \begin{cases} 1 & a = b \\ -\mu & a \neq b \end{cases} . \tag{5.122}$$

What is to be computed are the values of i, j and ℓ which lead to the highest total score

$$\Sigma \equiv \max_{A} S[A] \qquad (5.123)$$

for the given scoring matrix $s_{a,b}$.

This optimization problem can be reformulated by introducing the auxiliary quantity $S_{i,j}$ which is the optimal score of the subsequences ending at (i, j) optimized over ℓ. This quantity can be computed with $O(N^2)$ steps instead of $O(N^3)$ with the prescription

$$S_{i,j} = \max\{S_{i-1,j-1} + s_{a_i,b_j}, 0\} \qquad (5.124)$$

for the initial condition $S_{0,k} = S_{k,0} = 0$. This recursion expresses the fact that for a given pair (i, j) the optimal $\ell = 0$ or $\ell > 0$. If $\ell = 0$, the resulting score is zero either; if $\ell = 1$ at least, the corresponding pair (a_i, b_j) will belong to the optimal alignment (which may be longer), whatever had been chosen as optimal up to $(i - 1, j - 1)$. Eq. (5.124) describes a random walk with increment $s_{a,b}$. It is cut off when it falls below zero. The global score Σ is then given by

$$\Sigma = \max_{1 \le i \le M, 1 \le j \le N} S_{i,j} . \qquad (5.125)$$

Significance of the alignment. Suppose we have found an alignment by performing this computation. How significant is it? In order to answer this question, we have to discuss the alignment of purely random sequences, and to determine the distribution of scores in this case. This can be done rigorously, leading to the result

$$P[\Sigma < S] = \exp(-\kappa e^{-\lambda S}), \qquad (5.126)$$

which is a *Gumbel* or *extreme value distribution*. It is characterized by two parameters λ and κ where λ characterizes the tail of the distribution. For the case of gapless alignment we discuss here, both parameters can be calculated from the knowledge of the scoring matrix $s_{a,b}$.

We want to illustrate how the result (5.126) comes about in a heuristic fashion. For this we go back to Eq. (5.124) and set $i = j$, which is permissible for random sequences. Thus we have

$$S_{i,i} \equiv S(i) = \max\{S(i-1) + s(i), 0\} . \qquad (5.127)$$

In this equation, $s(i) = s_{a,b}$ plays the role of an uncorrelated noise given by the distribution

$$P[s(i) > s] = \sum_{(a,b|s_{a,b}>s)} p_a p_b . \tag{5.128}$$

Eq. (5.127) is essentially a discrete version of a Langevin equation. The dynamics generated can be found in two distinct phases, called the *linear* and the *logarithmic phase*. The quantity which distinguishes the two is the local similarity score

$$\langle s \rangle = \sum_{(a,b)} p_a p_b . \tag{5.129}$$

In the linear phase, the dynamics is a random walk with an average upward drift $\langle s \rangle$, and the maximal score is $\sum \approx N \langle s \rangle$ for a sequence of length N. This phase constitutes a phase of global alignment, and hence does not permit to identify similarities in subsequences. The distribution of the Σ in this phase is Gaussian, not Gumbel. If, however, the average drift $\langle s \rangle$ is negative, the ensuing dynamics will lead to a score of the form shown in Figure 5.3. The resulting

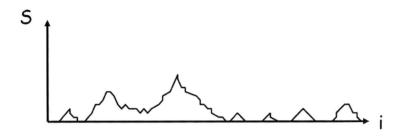

Figure 5.3 Total score of an alignment as a function of sequence position.

score landscape can be considered as consisting of 'islands' in an 'ocean'. For the case of random sequences, the islands are statistically independent. If σ_k is the maximal score of island k, the σ_k are thus independent random variables.

Island distribution. In order to calculate the island distribution explicitly, we have to look at (R. BUNDSCHUH, 2002)

$$p(\sigma) = \langle \delta(\sigma - \sum_{i=1}^{L} s(i)) \rangle . \tag{5.130}$$

Here L is the length of a 'large' island, measured from its beginning to the peak at height σ. Using the Fourier representation of the δ-function and the statistical independence of the $s(i)$ we obtain

$$p(\sigma) = \frac{1}{2\pi} \int dk e^{-ik\sigma} \langle e^{iks} \rangle^L \qquad (5.131)$$

Assuming that the peak score of each island is proportional to its length - islands are thus on average little triangles with rising slope α - we approximate

$$p(\sigma) \approx \frac{1}{2\pi} \int dk e^{-ik\alpha L} \langle e^{iks} \rangle^L \qquad (5.132)$$

and then evaluate the integral in a saddle-point approximation. This leads to (*Task!*)

$$p(\sigma) \sim \exp(-\lambda\sigma) \qquad (5.133)$$

with

$$\lambda = ik_s - \ln[\langle e^{ik_s s} \rangle]/\alpha, \qquad (5.134)$$

where the saddle-point value of k, which we call k_s, is determined by

$$\frac{\langle s e^{ik_s s} \rangle}{\langle e^{ik_s s} \rangle} = \alpha. \qquad (5.135)$$

Note that k_s still depends on α. The correct value α is found by minimizing Eq. (5.135) with respect to α and using Eq. (5.135), which yields

$$\langle e^{ik_s s} \rangle = 1 \qquad (5.136)$$

i.e., explicitly

$$\langle e^{\lambda s} \rangle = \sum_{a,b} p_a p_b e^{\lambda s_{a,b}} = 1. \qquad (5.137)$$

The typical slope of an island is given by

$$\alpha = \langle s e^{\lambda s} \rangle. \qquad (5.138)$$

Thus, we conclude that the islands follow the *exponential distribution*

$$P[\sigma_k > \sigma] \approx C e^{-\lambda\sigma}. \qquad (5.139)$$

Since the global optimal score Σ is given in terms of the islands σ_k by

$$\Sigma = \max_k \{\sigma_k\} \qquad (5.140)$$

the distribution of the Σ can be computed from the distribution of the σ_k.

For a large number $K \sim N$ of island peaks one finds

$$P[\Sigma < S] = P[\max\{\sigma_1, .., \sigma_K\} < S] = [1 - Ce^{-\lambda S}]^K \approx exp(-\kappa e^{-\lambda S}),$$
$$(5.141)$$

where $\kappa = CK$.

Gapped alignment. We now turn to the case of alignment with gaps. This method is used to detect weak sequence similarities between evolutionary distant sequences, in which deletions and insertions have occurred over time. The classic example is *Smith-Waterman local alignment*, in which the two subsequences $\mathbf{a}, \mathbf{b}$, e.g. GATGC and GCTC, can be aligned as GATGC and GCT-C, i.e., with one gap (see Figure 5.4). The score function for alignment with gaps is given by

$$S[A] = \sum_{a,b} s_{a,b} - \delta N_g,$$
$$(5.142)$$

where N_g is the number of gaps with cost δ.

The gapped alignment of two sequences can be represented as a directed path on a two-dimensional lattice, see Figure 5.4. The alignment score is the sum over local scores of the traversed bonds, whereby diagonal bonds are gaps with penalty δ, while horizontal bonds are given the *similarity scores* $s(r,t) \equiv s_{a_i,b_j}$. What is sought is the best scoring path connecting the lattice origin $(0,0)$ to its end, $(0, 2N)$.

If we denote by $h(r,t)$ the score of the best path ending in a lattice point $h(r,t)$, the highest scoring *global* alignment can be computed with the *Needleman-Wunsch algorithm*

$$h(r,t+1) = \max\{h(r,t-1) + s(r,t), h(r+1,t) - \delta, h(r-1,t) - \delta\}. \quad (5.143)$$

This expression can be interpreted to describe the configuration of a *directed polymer* in a random potential given by the local scores $s(r,t)$. Another interpretation, which is somewhat easier to visualize, is to understand the configuration as the height profile $h(r,t)$ of a growing interface between a solid and, say, a liquid phase.

These systems - *directed polymers* in a random potential or a *growing interface* - are well known in the physics of non-equilibrium systems. Due to the mapping of the sequence alignment problem onto the interface growth problem one can immediately show, from the knowledge of the models in non-equilibrium physics, that it belongs to the so-called *KPZ universality class*. A universality

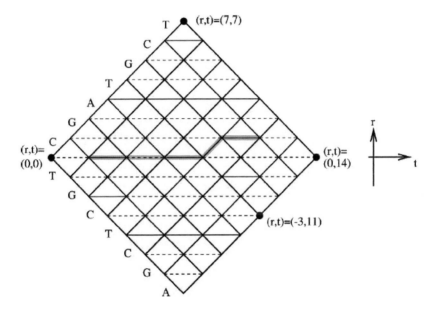

Figure 5.4 The local alignment of the two sequences CGATGCT and TGCTCGA is represented as a directed path on an alignment lattice: the diagonal bonds correspond to gaps, while horizontal bonds are aligned pairs. [Reprinted with permission from R. BUNDSCHUH (2002). Copyright: American Physical Society.]

class means that all systems that can be shown to be mathematically equivalent will have the same phase transition, i.e., it can be characterized by a common set of critical exponents.

Within a continuum description, one can show that the evolution of the heights is governed by the *Kardar-Parisi-Zhang equation* (often short: KPZ equation) (M. KARDAR, G. PARISI and Y.-C. ZHANG, 1986; T. HWA and M. LÄSSIG, 1998)

$$\partial_t h = \nu_0 \partial_r^2 h + \lambda_0 (\partial_r h)^2 + \eta(r, t) \tag{5.144}$$

where $\eta(r, t) = \frac{1}{2} s(r, t) - \delta$ is an uncorrelated Gaussian white noise. For $t \to \infty$ the heights assume a stationary equal-time distribution

$$P[h(r, t \to \infty)] \propto \exp\left(-\frac{1}{2D} \sum_r [h(r + 1, t) - h(r, t)]^2 \right). \tag{5.145}$$

The parameter D in this expression is a function of the scoring parameters μ and δ.

Typical score profiles $h(r, t)$ are illustrated in Figure 5.5 (top). In order to extract the characteristics of these profiles, it is instructive to consider the width of the profile, $w(t)$, defined by

$$w^2(t) = \frac{1}{X} \sum_{x=-X/2}^{x=X/2} [h(x, t) - h(t)]^2 \tag{5.146}$$

with the interval size $X (\approx N)$, and where the spatial average of the height, $h(t)$ is defined by

$$h(t) = \frac{1}{X} \sum_{x=-X/2}^{x=X/2} h(x, t). \tag{5.147}$$

The behaviour of h and w is shown in Figure 5.5 (bottom). As illustrated in the figure, the asymptotic behaviour of the width obeys a scaling law

$$\overline{w}(t) = B(\mu, \delta) t^\omega \tag{5.148}$$

where the value of ω is known from the universal scaling behaviour of the KPZ equation. Note that the scaling behaviour is obtained from an ensemble average over different realizations of random sequences. The quantity

$$\Delta h(t) = h_{max}(t) - h_{min}(t) \tag{5.149}$$

displays the same scaling behaviour as w.

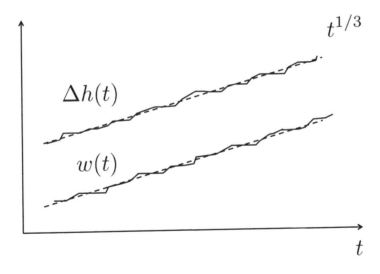

Figure 5.5 Qualitative sketch of the score profiles $h(x,t)$ (top) and scaling result for $h(t)$ and $w(t)$ (bottom).

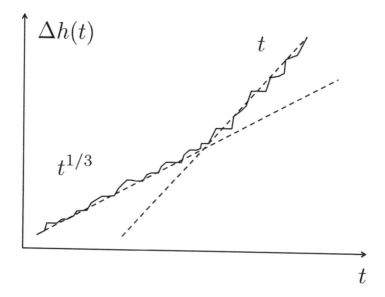

Figure 5.6 Qualitative sketch of the score profiles for weakly correlated sequences.

If the sequences exhibit correlations, the behaviour changes. This is illustrated in Figure 5.6, which displays the growth of a peak in the score profiles. In terms of the height difference $\Delta h(t)$ one sees a deviation from the scaling behaviour of the random case, from a power law of the form $t^{1/3}$ to a linear law. This change in exponent illustrates the onset of global alignment, and we are thus back to the linear phase we discussed before for the case of gapless alignment.

As a final step we want to locate the logarithmic phase in parameter space. For the gapped case, a negative drift $\langle s \rangle$ is not sufficient, since the average score now has to grow by a gap-dependent amount $u(\{s_{a,b}\}, \delta)$ on top of the expectation value $\langle s \rangle$. Consequently, the log-linear transition occurs at

$$u(\{s_{a,b}\}, \delta) + \langle s \rangle = 0. \tag{5.150}$$

In a (δ, μ) diagram this condition corresponds to a line $\delta_c(\mu)$, which can be calculated approximately or numerically. The result is shown in Figure 5.7.

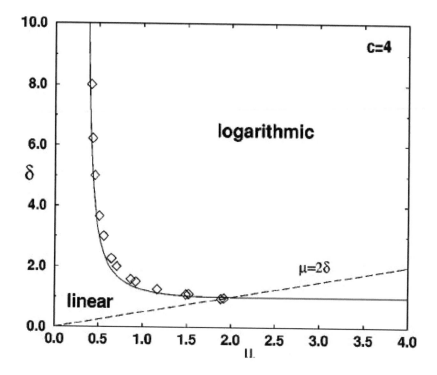

Figure 5.7 The phase diagram of the log-linear phase transition in the parameters δ vs. μ. The variable c the number of letters in the alphabet; a value of 4 hence refers to the alphabet of DNA. (Reprinted with permission from R. BUNDSCHUH, 2002. Copyright: American Physical Society.)

Additional Notes

Non-equilibrium processes. The basic reference for the description of non-equilibrium processes in physics and chemistry is the book by N. VAN KAMPEN (1992), which also served as a basis for the discussion presented here. Readers willing to learn more details are thus recommended to continue further studies there.

We have treated chemical kinetics in a very brief manner; for a much more detailed exposition, readers are asked to consult the book by PÉTER ÉRDI and GÁBOR LENTE (2014).

Stochastic thermodynamics. In recent years a novel approach has emerged for the discussion of thermodynamic processes in non-equilibrium. It is based on the Langevin-Fokker-Planck dichotomy. For a system in which temperature can be assumed to vary over time, and with the application of a deterministic force, described by a Langevin equation, the corresponding probability distribution P from the Fokker-Planck equation may be employed to define a non-equilibrium expression for entropy. This procedure generalizes equilibrium thermodynamics and is recovered for time-invariant states. Introductions are provided by K. SEKIMOTO'S book (2010) and U. SEIFERT'S review (2012).

Sequence alignment. The literature on sequence alignment is (fairly obviously) extensive. A point of reference is the book by M. S. WATERMAN (1995). The classic paper on alignment is by S. B. NEEDLEMAN AND C. D. WUNSCH (1970). For a detailed description of the development until 1995 the reader is asked to consult the *Sources and Perspectives* section in Waterman's book.

The relationship between the alignment problem and models from non-equilibrium statistical mechanics was observed by T. HWA AND M. LÄSSIG (1996, 1998). More recently, a mapping to the *asymmetric exclusion process* has been proposed by R. BUNDSCHUH (2002) which is the same universality class as the KPZ-equation discussed in the text. The Asymmetric Exclusion Process (ASEP) is a favorite model for studies of non-equilibrium statistical mechanics since it is amenable to rigorous approaches.

The literature on both the ASEP and in recent years in particular on the KPZ-equation has literally exploded, in particular through the mathematical advances by M. HAIRER on the mathematical theory of stochastic differential equations. The literature in the field is enormous; a recent paper permitting a self-contained introduction to the topic is by I. CORWIN (2018).

References

R. Bundschuh
Asymmetric exclusion process and extremal statistics of random sequences
Phys. Rev. E **65**, 031911 (2002)

I. Corwin
Exactly solving the KPZ equation
in: Random Growth Models, Proceedings of Symposia in Applied Mathematics, **75**, 203-254 (2018)

P. Érdi, G. Lente
Stochastic Chemical Kinetics: Theory and (Mostly) Systems Biological Applications
Springer (2014)

D.T. Gillespie
Exact stochastic simulation of coupled chemical reactions
J. Phys. Chem. **81**, 2340-2361 (1977)

T. Hwa, M. Lässig
Similarity detection and localization
Phys. Rev. Lett. **76**, 2591-2594 (1996)

T. Hwa, M. Lässig
Optimal detection of sequence similarity by local alignment
in *Proceedings of the Second Annual Conference on Computational Molecular Biology*, S. Istrail (ed.), ACM Press (1998)

M. Kardar, G. Parisi, Y.-C. Zhang
Dynamic scaling of growing interfaces
Phys. Rev. Lett. **56**, 889-892 (1986)

T. B. Liverpool
Non-equilibrium systems have steady-state distributions and non-steady dynamics
arXiv:1810.10980 cond-mat.soft

C. Merlo, K.A. Dill, T.R. Weikl
Φ-*values in protein-folding kinetics have energetic and structural components*
Proc. Natl. Acad. Sci. USA **102**, 10171-10175 (2005)

S.B. Needleman, C.D. Wunsch
A general method applicable to the search for similarities in the amino acid sequence of two proteins
J. Mol. Biol. **48**, 443-453 (1970)

N.G. van Kampen
Stochastic Processes in Physics and Chemistry
Elsevier (1992)

U. Seifert
Stochastic thermodynamics, fluctuation theorems, and molecular machines
Rep. Prog. Phys. **75**, 126001 (2012)

K. Sekimoto
Stochastic Energetics
Springer, Heidelberg (2010)

M. Waterman
Introduction to Computational Biology
Chapman & Hall/ CRC (1995)

CHAPTER 6

Fluctuation Theorems

Non-equilibrium statistical mechanics still lacks general concepts when compared to equilibrium statistical mechanics. That does not say, however, that there are no fundamental relations between physical quantities that can be usefully formulated. The most important are *fluctuation theorems* which we will now introduce.

6.1 THE FLUCTUATION-DISSIPATION THEOREM

We first derive a general fluctuation theorem, valid for non-equilibrium stationary states (and, a fortiori, near equilibrium states) within the formalism we have introduced (M. LAX, 1960).

Suppose we have I molecular species present with numbers n_i, $1 \leq i \leq I$; we consider the index i as a vector index on particle numbers, and abbreviate $\mathbf{n} = (n_1, ... n_I) = n_i$, which is hence a row vector.

From our previous results we infer that the particle distribution function for a Markov process fulfills

$$P(\mathbf{n}, t + \Delta t) = \int d\mathbf{m} P(\mathbf{n}, t + \Delta t | \mathbf{m}, t) P(\mathbf{m}|t), \qquad (6.1)$$

We now define the following quantities[1]

$$\mathbf{A}(\mathbf{m}) = \frac{1}{\Delta t} \int d\mathbf{n} P(\mathbf{n}, t + \Delta t | \mathbf{m}, t)(\mathbf{n} - \mathbf{m}), \qquad (6.2)$$

---

[1] This step assumes that the conditional probabilities are expandable to linear order in Δt.

which is a vector; we call it a *drift vector*. Further, we define the matrix

$$\mathbf{D}(m) = \frac{1}{2\Delta t} \int dn P(\mathbf{n}, t + \Delta t | \mathbf{m}, t)(\mathbf{n} - \mathbf{m}) \cdot (\mathbf{n} - \mathbf{m})^T \qquad (6.3)$$

where $(\dots)^T$ denotes the transpose of the row vector of particle numbers, hence a column-vector. We call $\mathbf{D}$ the *diffusion matrix*.

Supposing that the *stationary state* is characterized by a particle number vector $\mathbf{n}_0$, we introduce the vector

$$\delta\mathbf{n} \equiv \mathbf{n} - \mathbf{n}_0 . \qquad (6.4)$$

We now find (verify this as a *Task*) the equation for the time evolution of the mean-value

$$\frac{d}{dt}\langle \delta\mathbf{n} \rangle = \langle \mathbf{A}(\mathbf{n}(t)) \rangle \qquad (6.5)$$

and the corresponding equation for the covariance matrix

$$\frac{d}{dt}\langle \delta\mathbf{n}\,\delta\mathbf{n}^T \rangle = 2\langle \mathbf{D}(\mathbf{n}) \rangle + \langle \mathbf{A}(\mathbf{n})\delta\mathbf{n}^T \rangle + \langle \delta\mathbf{n}\mathbf{A}^T(\mathbf{n}) \rangle . \qquad (6.6)$$

As a following step we expand around the stationary state via $\mathbf{n} = \mathbf{n}_0 + \delta\mathbf{n}$ and assume for the drift-vector

$$\mathbf{A}(\mathbf{n}) \approx \mathbf{A}(\mathbf{n}_0) - \mathbf{\Lambda}\delta\mathbf{n} , \qquad (6.7)$$

where we have introduced a matrix $\mathbf{\Lambda}$. In this last equation we choose

$$\mathbf{A}(\mathbf{n}_0) = 0 \qquad (6.8)$$

to be consistent with Eq. (6.5). Further, we make the simplifying assumption that

$$\mathbf{D}(\mathbf{n}) \approx \mathbf{D}(\mathbf{n}_0) \equiv \mathbf{D} . \qquad (6.9)$$

We then end up with the following results for the mean

$$\frac{d}{dt}\langle \delta\mathbf{n} \rangle = -\mathbf{\Lambda}\langle \delta\mathbf{n} \rangle \qquad (6.10)$$

and the *covariance matrix*

$$\frac{d}{dt}\langle \delta\mathbf{n}\delta\mathbf{n}^T \rangle = 2\mathbf{D} - \mathbf{\Lambda}\langle \delta\mathbf{n}\delta\mathbf{n}^T \rangle - \langle \delta\mathbf{n}\delta\mathbf{n}^T \rangle\mathbf{\Lambda}^T . \qquad (6.11)$$

These equations are the main result of this section. We note that in steady-state, a relation between the diffusion matrix $\mathbf{D}$, the fluctuations in particle number $\delta\mathbf{n}$ and the drift matrix $\boldsymbol{\Lambda}$ is established in the form

$$2\mathbf{D} = \boldsymbol{\Lambda}\langle\delta\mathbf{n}\delta\mathbf{n}^T\rangle + \langle\delta\mathbf{n}\delta\mathbf{n}^T\rangle\boldsymbol{\Lambda}^T . \tag{6.12}$$

This equation is a general version of the so-called *fluctuation-dissipation theorem*. The name derives from special cases in which the drift vector corresponds, e.g., to a friction force acting on a particle and hence provides a mechanism for dissipation. We have seen such situations already in the previous chapter, e.g. in the simple cases of Langevin dynamics.

Exercise. Apply the above theorem to the original Langevin equation discussed in Chapter 5. What do you conclude?

6.2   THE JARZYNSKI EQUALITY AND CROOKS' THEOREM

We now turn to another general result, the so-called *Jarzynski equality*. This time we start the derivation from the master equation, following U. SEIFERT, 2004.

We consider a situation in which we allow the w_{mn} in the master equation to depend on some tunable parameter λ, i.e., we have

$$w_{mn} = w_{mn}(\lambda) . \tag{6.13}$$

For a fixed value of λ, we assume (as before) that the system is in a stationary state p_n^s which obeys the condition of *detailed balance*. This condition is quite important: while obeyed by equilibrium systems, the reverse it not true. The condition of detailed balance suffices for the system to have a stationary state, and hence is an important information to have on a non-equilibrium system. Mathematically, the condition of detailed balance is given by

$$\frac{p_n^s}{p_m^s} = \frac{w_{mn}}{w_{nm}} . \tag{6.14}$$

If we now assume that the parameter λ is turned on in a time-dependent manner, $\lambda = \lambda(\tau)$, we would like to know the probability P to encounter a particular trajectory of the system $n(\tau) \equiv (n_0, n_1, ..., n_k)$ starting in n_0 at time $\tau_0 = 0$, jumping to n_1 after a time-interval τ_1 and so forth, until the final jump from τ_{k-1} to $\tau_k \equiv t$. This probability is given by

$$P[n(\tau), \lambda(\tau)] = p^s(n_0, \lambda(0)) \times \prod_{i=1}^{k} \exp\left[-\int_{\tau_i}^{\tau_{i+1}} d\tau \sum_{m \neq n_i} w_{n_0,m}(\lambda(\tau))\right]$$

$$\times w_{n_0,m}(\lambda_{\tau_{i+1}}).$$

(6.15)

Likewise, we can study the trajectory $\tilde{n} \equiv n(t-\tau)$ which occurs under reversal of λ, i.e., $\lambda(t-\tau) \equiv \tilde{\lambda}$. This operation allows to write down the probability $P[\tilde{n}(\tau), \tilde{\lambda}(\tau)]$ (*Exercise*).

We can then form the ratio of the two probabilities which is given by

$$e^{-R[n(t)]} \equiv \frac{P[\tilde{n}(\tau), \tilde{\lambda}(\tau)]}{P[n(\tau), \lambda(\tau)]} = \exp\left[-\int_0^t d\tau \epsilon'_{n(\tau)}\dot{\lambda}(\tau)\right]$$

(6.16)

where the quantity

$$\epsilon'_n(\lambda) = -\frac{d}{d\lambda} \ln p^s(n, \lambda)$$

(6.17)

has been introduced. It can be considered as a formal 'energy level'. This interpretation can be made if one wishes to read the integral in the argument of the exponential function as an analogue of a free energy - we will soon see that this analogy can be made precise.

From these observations we obtain the following two identities

$$1 = \sum_{\tilde{n}(\tau)} P[\tilde{n}(\tau), \tilde{\lambda}(\tau)] = \sum_{\tilde{n}(\tau)} e^{-R[n(\tau)]} P[n(\tau), \lambda(\tau)]$$

(6.18)

and

$$1 = \sum_{n(\tau)} e^{-R[n(\tau)]} P[n(\tau), \lambda(\tau)] = \left\langle \exp\left(-\int_0^t d\tau \epsilon'_{n(\tau)}\dot{\lambda}(\tau)\right)\right\rangle.$$

(6.19)

We interpret the second expression as a *fluctuation theorem*

$$\left\langle \exp\left(-\int_0^t d\tau \epsilon'_{n(\tau)}\dot{\lambda}(\tau)\right)\right\rangle = 1,$$

(6.20)

and explain now why, by some illustrative applications.

6.3   APPLICATIONS OF THE FLUCTUATION THEOREMS

We first illustrate the result Eq. (6.20) by applying it to a simple cyclically working enzyme or motor, following U. SEIFERT, 2005, as depicted in Figure 6.1. The enzyme is assumed to have three equivalent conformational states, and it progresses from one to the other at a rate k_+ in the forward (i.e., clockwise), and with a rate k_- in the backward direction. We assume $k_+ > k_-$. The stationary distribution of the system clearly is given by $p^s = 1/3$: the system spends equal times in each of the states. For this system we obtain R as

$$R = n \ln(k_+/k_-) \tag{6.21}$$

where $n \equiv n_+ - n_-$ is the effective number of steps in the forward direction. Thus

$$\frac{P[-n]}{P[n]} = e^{-n \ln(k_+/k_-)} = \left(\frac{k_+}{k_-}\right)^n . \tag{6.22}$$

The exact $P[n]$ can be computed from the master equation, since this system is an *asymmetric random walk* with the master equation

$$\dot{p}_n = k_+ p_{n+1} - k_- p_{n-1} - (k_+ + k_-)p_n \tag{6.23}$$

for which

$$p_n \equiv P[n] = I_{|n|}(2\sqrt{k_+k_-}t)\left(\frac{k^+}{k^-}\right)^{n/2} e^{-(k_+ + k_-)t}, \tag{6.24}$$

where $I_n(x)$ is the modified Bessel function of order[2] n. One sees that the factor $(k_+/k_-)^n$ arises from the ratio of backward and forward processes with the probability ratio $P[-n]/P[n]$.

Jarzynszki theorem. The fluctuation theorem above can further be elucidated by pointing out its relation of Eq. (6.20) with a theorem due to C. JARZYNSKI, 1997a,b, as suggested by U. SEIFERT, 2004.

---

[2]The modified Bessel function of order n is given by the expression

$$I_\alpha(x) = i^{-\alpha} J_\alpha(x)$$

with

$$J_\alpha(x) = \sum_{m=0}^{\infty} \frac{(-1)^m}{m!\Gamma(m + \alpha + 1)} \left(\frac{x}{2}\right)^{2m+\alpha} .$$

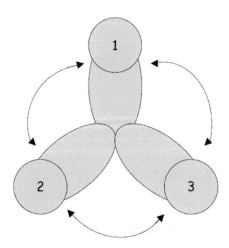

Figure 6.1 An enzyme or molecular motor switching between three configurational states. Transition rates in forward and backward direction are given by k_i^+ and k_i^- for $i = 1, 2, 3$, respectively.

Consider the trajectory of a particle $x(t)$ in a potential $V(x, \lambda)$, which again depends on the parameter λ. The stationary distribution of this process for fixed λ is given by

$$p^s(x, \lambda) = Z(\lambda)^{-1} \exp[-\beta V(x, \lambda)] \tag{6.25}$$

with the normalization given by the 'partition function'

$$Z(\lambda) \equiv \int_{-\infty}^{\infty} dx \, \exp[-V(x, \lambda)] \,. \tag{6.26}$$

Substituting these correspondences into the formula (6.17) one has with $n(t) \sim x(t)$

$$\epsilon_n(\lambda) \sim -\ln p^s(x, \lambda) = \beta V(x, \lambda) + \ln Z(\lambda) \,. \tag{6.27}$$

Inserted into Eq. (6.20) this leads to the result

$$\langle \exp\left(-\beta \int_0^t d\tau V'(x(\tau), \dot{\lambda}(\tau))\right) \frac{Z(\lambda(0))}{Z(\lambda(t))} \rangle = 1 \,. \tag{6.28}$$

Invoking the expression for the free energy

$$F(\lambda) = \beta^{-1} \ln Z(\lambda), \tag{6.29}$$

and introducing the notions of the *applied work*

$$W \equiv \int_0^t d\tau V'[x(\tau), \lambda(\tau)]\dot{\lambda}(\tau) \tag{6.30}$$

and the *dissipated work*

$$W_{diss} \equiv W - [F(\lambda(t)) - F(\lambda(0))] \tag{6.31}$$

we finally arrive at the expression

$$\langle e^{-\beta W_d} \rangle = 1 \tag{6.32}$$

or, equivalently, at

$$\langle e^{-\beta W} \rangle = e^{-\beta \Delta F} \tag{6.33}$$

where ΔF can be read off from Eq. (6.31). This is Jarzynski's original result (1997a).

The Jarzynski equation (6.33) is remarkable since it relates the difference between an *equilibrium free-energy difference* to the average over a *non-equilibrium quantity*. This merits a deeper discussion.

Validation of Jarzynski's equality. Jarzynski's result can be rewritten as an equation for the free energy difference in terms of a non-equilibrium expression

$$\Delta F = \beta^{-1} \ln\langle e^{-\beta W} \rangle. \tag{6.34}$$

To interpret this result further it is useful to rewrite it further in terms of the *cumulant expansion* (see Chapter 1). This operation leads to the expression

$$\Delta F = \sum_{n=1}^{\infty} \frac{1}{n!} \kappa_n (-\beta)^{n-1} \tag{6.35}$$

where the first four cumulants are given by

$$\kappa_1 = \langle W \rangle \ , \quad \kappa_2 = \langle (W - \langle W \rangle)^2 \rangle = \sigma_W^2 \tag{6.36}$$

and

$$\kappa_3 = \langle (W - \langle W \rangle)^3 \rangle \ , \quad \kappa_4 = \langle (W - \langle W \rangle)^4 \rangle - 3\sigma_W^4 \ . \tag{6.37}$$

These expressions are instructive since we can infer from them different levels of approximation which are testable against experimental measurements or simulation results.

To begin, we keep only the first term of the cumulant expansion. We then estimate the work done by the system as the free energy difference. This is correct only when the work is done *reversibly*, which means that there is no mechanism of energy dissipation in the system.

The second level approximation amounts to consider

$$\Delta F = \langle W \rangle - \frac{1}{2}\beta\sigma_W^2 , \tag{6.38}$$

i.e., the fluctuations around the stationary state. Eq. (6.38) becomes exact in a regime near an equilibrium state, since then the work distribution is Gaussian, and all higher cumulants vanish identically. The result is also an example application of the *fluctuation-dissipation theorem*, since

$$\overline{W}_{diss} = \frac{1}{2}\beta\sigma_W^2 , \tag{6.39}$$

relates the dissipated work to the Gaussian fluctuations.

If we want to take the full Jarzynski result serious, we have to estimate the free energy difference by the following expression

$$\Delta F = -\frac{1}{\beta}\ln\left[\frac{1}{N}\sum_{i=1}^{N}e^{-\beta W_i}\right] , \tag{6.40}$$

where N is the number of trajectories for which the work W has been determined.

What trajectories are best to measure in an experiment or a simulation? This can be understood from reconsidering the Jarzynski result in the form involving the dissipated work. Since we are near an equilibrium state, W is Gaussian-distributed, hence the dissipated work follows a Gaussian distribution with mean $\overline{W}_{diss} = \frac{1}{2}\beta\sigma_W^2$ and variance σ_W^2. As $\overline{W}_{diss}$ increases - one moves away from the near-equilibrium regime - the distribution will broaden. The expression Eq. (6.40) obviously heavily weighs those trajectories whose dissipated work is *negative*. The probability of finding such a trajectory is given by

$$P(W_{diss} < 0) = \int_{-\infty}^{0} dW_{diss}P(W_{diss}) = \frac{1}{2}(1 - \mathrm{erf}(\sqrt{\overline{W}_{diss}}/2)) \tag{6.41}$$

where erf is the *error function*.[3] Since Eq. (6.41) is a sharply decreasing function of its argument, it is established that the efficiency of sampling falls rapidly with the increase of dissipated work.

Pulling RNA and DNA hairpins. Jarzynski's equality has been tested in experiments on pulling RNA secondary structure. In such experiments - using a similar experimental setup we described in Chapter 2 of Part I in the context of the worm-like chain model - a single RNA molecule with known secondary structure is fixed at its extremities between two beads which can be moved reversibly, at different speeds (J. LIPHARDT et al., 2002). These experiments have indeed allowed to demonstrate that the estimator given by Eq. (6.40) converges towards ΔF if sufficiently many trajectories N are taken into account for which the dissipated work is high. The quantity can also be validated in simulations (S. PARK et al., 2003). Again, a sufficient sampling range is needed. More recently, a detailed study of RNA folding free energies has been performed by D. COLLIN et al., 2005, who verified a generalization of the Jarzynski theorem, the *Crooks fluctuation theorem*, in regimes near and far from equilibrium. In the context of RNA folding, this theorem states that the probability the ratio of the probability distributions of the work for unfolding and folding under conditions of time-reversal symmetry fulfills

$$\frac{P_{unfold}(W)}{P_{fold}(W)} = \exp \beta (W - \Delta F) \,. \tag{6.42}$$

In later work, the pulling of DNA hairpins was investigated in order to further quantify the conditions under which the Jarzynski result holds, as there are numerous sources of error when comparing real experiments with the basic theoretical insight expressed in Jarzynski's result. We now analyze the case of DNA hairpins in somewhat more detail, trying to find out in what variables the measured work has to be defined. We follow the discussions in J. GORE et al. (2013) and A.M. MONGE et al. (2018).

The setup of the experiment that we consider is shown in Figure 6.2. One of the beads, on which the hairpin extremities are fixed, is held by a pipette in a fixed position (0). The second bead has the position $x(t)$, as it moves away under the influence of the trap with velocity v. The trap itself is centered at $\lambda(t) > x(t)$. The experimental control parameter is this length $\lambda(t)$, while the molecular extension, the fluctuating quantity, is $x(t)$. The DNA hairpin is a pure two-state system (see Chapter 2), either fully open or fully closed (except for a small loop at the top). Both the trap and the hairpin molecule

---

[3]The *error function* is defined by

$$\mathrm{erf}(x) \equiv \frac{2}{\sqrt{\pi}} \int_0^x dt e^{-t^2} \,.$$

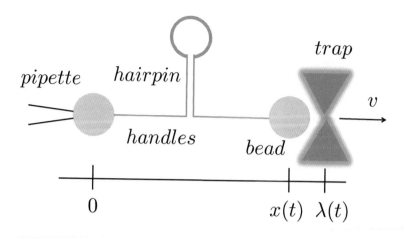

Figure 6.2 Schematic DNA hairpin experiment. (After A. MOSSA et al., 2009.)

+ handles, which fix the hairpin are considered as two harmonic potentials with stiffnesses k_b and k_m. The rest length of the trap spring is zero, while the rest length for the molecular spring k_m is either ℓ_0 if the hairpin is closed, or ℓ_1, if the hairpin is open. For each value of λ, the state of the system is then defined by a pair (x, a) where $a = 0$ denotes closed hairpin, and $a = 1$ denotes open hairpin. The force acting on the right bead is thus given by

$$f(x, a) = k_b(\lambda - x) - k_m(x - \ell_a). \tag{6.43}$$

which can be rewritten as

$$f(x, a) = -k_t(x - x_{eq}), \tag{6.44}$$

where x_{eq} is the equilibrium position

$$x_{eq} = \frac{k_b\lambda + k_m\ell_a}{k_t} \tag{6.45}$$

with the total stiffness $k_t = k_b + k_m$.

We further can assume mechanical equilibrium, $\langle f_t(t) \rangle = 0$, and write down the Hamiltonian corresponding to the force as

$$H^{(\lambda)} = \frac{k_b}{2}(\lambda - x)^2 + \frac{1}{2}(x - \ell_a)^2 + a\Delta G_0, \tag{6.46}$$

where ΔG_0 is the free energy difference between the open and closed states of the hairpin without applied force.

Task. Calculate the partition function for the Hamiltonian (6.46) and derive expressions for the expectation values of the molecular extension, $\langle x \rangle$, and the variance of x.

For the kinetics we assume rates of opening and closing according to a *Bell-Kramers type barrier crossing*, i.e. exponentials reading as

$$k_\rightarrow = k_0 \exp\left(\frac{w_0 f_0(x)}{k_B T} \right) \tag{6.47}$$

and

$$k_\leftarrow = k_0 \exp\left(\frac{-w_1 f_1(x) + \Delta G_0}{k_B T} \right) \tag{6.48}$$

with w_i as distances from the barrier to closed and open states, f_i two forces, and k_0 as a typical *attempt frequency*. To derive the relations between these quantities we enforce detailed balance,

$$\frac{k_\rightarrow}{k_\leftarrow} = \exp\left[-\frac{H^{(\lambda)}(x,1) - H^{(\lambda)}(x,0)}{k_B T} \right] \tag{6.49}$$

which holds for all λ and x. Relating exponentials we obtain for the simplifying choice of $f_0(x) = f_1(x)$ the remaining condition

$$w_0 + w_i = \ell_1 - \ell_0. \tag{6.50}$$

The dynamics of the model is given by the Langevin equation

$$\gamma \dot{x} = f_t(x(t), a) + \sqrt{2\gamma k_B T} \xi(t) \tag{6.51}$$

with γ as the *friction coefficient* of the bead in the trap, and $x(t)$ as a Gaussian white noise (see Chapter 5). Finally, the experimental protocol is defined by the imposed function $\lambda(t) = \lambda_0 + vt$.

We can now turn to discussing the work associated with the induced transition between the open and closed hairpin structure. In the course of a pulling experiment Γ occurring between an initial point i and a final point f one has the integral over the force along the trajectory

$$W_\Gamma = \int_{\lambda_i}^{\lambda_f} d\lambda f_b(\lambda, x) \tag{6.52}$$

with the force induced by the displacement of the trapped bead. However, this is not the work obtained in the pulling experiment, which is instead given by

$$W'_\Gamma = \int_{x_i}^{x_f} dx f(\lambda, x) \tag{6.53}$$

along the actual trajectory of the experiment Γ, to which correspond the molecular extensions at time t_i and $t_f = t_i + (\lambda_f - \lambda_i)/v$, which are given by

$$x = \lambda - \frac{f}{k_b}. \tag{6.54}$$

Therefore, the two work expressions are related to each other via

$$W_\Gamma = W'_\Gamma + \frac{f_f^2 - f_i^2}{2k_b}. \tag{6.55}$$

The correction term only corresponds to a contribution from the boundaries of the experiment.

If we now hope to realize an experiment in quasi-equilibrium conditions, for pulling speeds $v \to 0$, we find two expressions for the free energy difference ΔG_0 we are interested in. We have

$$\Delta G_0 = W_{rev} - \frac{\langle f_f \rangle^2 - \langle f_i \rangle^2}{2k_{\text{eff}}} \tag{6.56}$$

where

$$\frac{1}{k_{\text{eff}}} = \frac{1}{k_b} + \frac{1}{k_m}. \tag{6.57}$$

If we plot the force-extension curve as a function of molecular extension, we have instead

$$\Delta G_0 = W'_{rev} - \frac{\langle f_f \rangle^2 - \langle f_i \rangle^2}{2k_m}. \tag{6.58}$$

We can now finally turn to the application of the Jarzynski equality, from which we can estimate the reversible work, and this is indeed the quantity W, while the experimenter often has W' at his disposal. We can evaluate the error that occurs when the proper one is replaced by the alternative quantity. For this we consider the reversible work measured in n experiments and used to calculate the average

$$\beta \tilde{W} = -\ln \sum_{i=1}^{n} \frac{1}{n} \exp(-\beta W_i), \tag{6.59}$$

and likewise for W'. To estimate the difference between $\tilde{W}$ and $\tilde{W}'$, we can sort the measured work function according to their magnitudes. The sum of exponentials is dominated by the minimum-work trajectory,

$$\beta(\tilde{W}) \approx \beta\tilde{W} + \ln n \,, \tag{6.60}$$

which also holds for $\tilde{W}'$. Therefore the difference in work is given by the difference in minimal work trajectories. We thus need to know the distribution of the trajectories. It turns out that the trajectories for W' are Gaussian-distributed, while those for W follow approximately a Gumbel distribution, which is slightly skewed when compared to a Gaussian. One obtains (MOSSA et al., 2009) approximate difference between the estimators,

$$\Delta G_0 - \Delta G_0' \approx \frac{\sqrt{6}}{\pi}(\gamma - \ln n)s + \sqrt{2}z(n)s' \tag{6.61}$$

where one has identified $W_{rev} = \tilde{W}$ and defined $\Delta G_0'$ via the imposition of $W'_{rev} = \tilde{W}'$. In Eq. (6.61), s, s' are the standard deviations of the distributions of W_i and W_i', γ is the Euler-Mascheroni constant and $z(n)$ is the solution of a transcendental equation,

$$\sqrt{\pi}z[1 + \mathrm{erf}(z)] = (n-1)\exp(-z^2) \,, \tag{6.62}$$

which essentially grows logarithmically in n.

Additional Notes

The relation between equilibrium and non-equilibrium physics as expressed by the Jarzynski theorem has been an active field of research ever since its inception. In particular the recent papers by C. JARZYNSKI, G. CROOKS, C. MAES and U. SEIFERT are recommended reading for those who want to know more on this topic.

On the experimental side, there are continuous efforts to validate the fluctuation theorems. Three earlier contributions in this field are by D. KELLER et al. (2003), E. H. TREPAGNIER et al. (2004), and D. COLLIN et al. (2005). A nice summary and discussion of the developments in the field is by C. JARZYNSKI, 2010.

References

D. Collin, F. Ritort, C. Jarzynski, S.B. Smith, I. Tinoco Jr., C. Bustamante
Verification of the Crooks fluctuation theorem and recovery of RNA folding free energies
Nature **437**, 231-234 (2005)

G.E. Crooks
Path-ensemble averages in systems driven far from equilibrium
Phys. Rev. E **61**, 2361-2366 (2000)

J. Gore, F. Ritort, C. Bustamante
Bias and error in estimates of equilibrium free-energy differences from nonequilibrium measurements
Proc. Natl. Acad. Sci. USA **100**, 12564-12569 (2003)

C. Jarzynski
Nonequilibrium equality for free energy differences
Phys. Rev. Lett. **78**, 2690-2693 (1997a)

C. Jarzynski
Equilibrium free-energy differences from nonequilibrium measurements: A master-equation approach
Phys. Rev. E **56**, 5018-5035 (1997b)

C. Jarzynski
Equalities and inequalities: irreversibility and the second law of thermodynamics at the nanoscale
Séminaire Poincaré XV Le Temps, 77-102 (2010)

D. Keller, D. Swigon, C. Bustamante
Relating single-molecule measurements to thermodynamics
Biophys. J. **84**, 733-738 (2003)

M. Lax
Fluctuations from the nonequilibrium steady state
Rev. Mod. Phys. **32**, 25-47 (1960)

J. Liphardt, S. Dumont, S.B. Smith, I. Tinoco jr., C. Bustamante
Equilibrium information from nonequilibrium measurements in an experimental test of Jarzynski's equality
Science **296**, 1832-1835 (2002)

C. Maes
On the origin and use of fluctuation relations for the entropy
Sém. Poincaré **2**, 29-62 (2003)

A.M. Monge, M. Manosas, F. Ritort
Experimental test of ensemble inequivalence and the fluctuation theorem in the force ensemble in DNA pulling experiments
Phys. Rev. E **98**, 032146 (2018)

A. Mossa, S. de Lorenzo, J.M. Huhuet, F. Ritort
Measurement of work in single-molecule pulling experiments
J. Chem. Phys. **130**, 234116 (2009)

S. Park, F. Khalili-Araghi, E. Tajkhorshid, K. Schulten
Free energy calculation from steered molecular dynamics simulations using Jarzynski's equality
J. Chem. Phys. **119**, 3559-3566 (2003)

U. Seifert
Fluctuation theorem for birth-death or chemical master equations with time-dependent rates
J. Phys. A: Math. Gen. **37**, L517-L521 (2004)

U. Seifert
Fluctuation theorem for a single enzym or molecular motor
Europhys. Lett. **70**, 36-41 (2005)

E.H. Trepagnier, C. Jarzynski, F. Ritort, G.E. Crooks, C.J. Bustamante, J. Liphardt
Experimental test of Hatano and Sasa's nonequilibrium steady-state equality
Proc. Natl. Acd. Sci. USA **101**, 15038-15041 (2004)

CHAPTER 7

Dynamics of Biological Networks

How come all my body parts so nicely fit together?
All my organs doing their jobs, no help from me!

Crash Test Dummies, "How Does a Duck Know?" (1993)

The detailed knowledge of the properties of biomolecules and their interactions, as we have described them in the previous chapters is a necessary but not a sufficient step to understand biological systems. The next step is to understand how the different molecular components *interact in concert* and how they build up the hierarchy observed in biological systems: cellular compartments, cells, organs and organisms.

In most of what follows, we will be interested here in *genetic networks* of some sort. Under this term we understand biological reaction schemes on the basis of the central dogma DNA makes RNA makes protein. We will encounter the following situations:

- A protein attaches itself to DNA at a binding site; such molecules are *transcription factors* which help control gene expression;

- A protein-RNA complex, the RNA polymerase, reads out the genes after fixing itself at a *promoter site*; the readout is a messenger RNA;

- The mRNA transcript is read at a ribosome and serves as a blueprint for a protein;

- Proteins compete with each other; this may happen by their direct interaction or blocking or activating the corresponding genes.

This basic readout mechanism underlies the complexity of all gene networks. The examples we discuss have been selected to allow for, mostly, analytical calculations to illustrate several aspects arising in *systems biology*.

7.1 THE λ-REPRESSOR

In this section we want to discuss an example of a biological network - a very small one - and to formulate a simple model for a particular aspect of its dynamics. The example is the λ-*phage*.

The λ-phage is a bacterial virus for which the molecular basis of its lifecycle is very well understood (M. PTASHNE, 2004). Consequently, it is often used as a prototypical example for theoretical studies. Here, we use it as a 'simple' complex system to illustrate how to describe biological networks within a deterministic dynamics setting. But first, we need to learn some basics of the lifecycle of the virus; many more details can be found in Ptashne's book.

Lifecycle of phage λ. Phage λ is a bacterial virus which infects *E. coli*. The virus inserts its DNA into the bacterial cell. Here, the DNA has two options. The first is its insertion into the host genome, and it then 'stays on board' as a 'blind passenger', being passed on from one generation to the next upon cell division. This phase is called the *lysogenic phase*. But if, e.g., the bacterium is endangered and responds to this external stress, the virus can 'abandon ship'. It genome is excised out of the host genome and begins the production of new viruses which then leave the host (which dies). The phase in which the virus replicates is called the *lytic phase*.

The decision making process between the lysogenic or lytic phase is made at a molecular switch. The switch is illustrated schematically in Figure 7.1. It consists of an *operon* built from three *operator binding sites*, at which a dimer of the λ-repressor molecule can attach itself. The λ-repressor is a *transcription factor*, i.e., its presence on DNA serves as a flag for RNA polymerase to attach itself and start transcription of an adjacent gene. Indeed, the three operator sites are at the same time overlapping with two neighbouring *promoter* regions, which are the fixation platforms for RNA polymerase, and they directly neighbour two adjacent genes, *cI*, the gene for the λ-repressor molecule, and *cro*, the gene for a second transcription factor also involved in the switch. Although operator and promoter sites are overlapping - and part of their function depends on this fact - their biological role is different.

Figure 7.1 (**See color insert.**) The λ-phage operon. Operator sites are shown as gray-shaded areas, with OR3-OR2-OR1 from left to right. Promoters are indicated by dashed frames; gene transcription directions are indicated by arrows. Two λ-repressor dimers are shown upon attachment and approach.

The switch now functions as follows. The attachment of a cI-dimer (the other name of the dimer form of the λ-repressor) at operator site OR1 enhances the fixation probability of a second dimer to fix at OR2. The fixation of cI at OR2, in turn, increases the probability of fixation of RNA polymerase at the cI-promoter, upon which transcription of cI occurs. The action of cI is thus directly implied in its description, hence the gene is *autoregulatory*. The continual transcription of cI ensures that the system stays in the lysogenic phase.

If, on the other hand, the system needs to switch to the lytic phase, a repressor dimer can fix at OR3, which due to its overlap with the cI-promoter blocks the access for RNA-polymerase. Now, Cro-proteins intervene, and they attach themselves to the same OR-sites as CI, but with inverted affinities. A Cro-protein present at OR2 enhances the fixation probability for a polymerase at the cro-promoter, and transcription of cro can start, initiating the lytic pathway.

Let's now build some of these mechanisms into a mathematical model.

Hasty model I: basics. In this subsection we discuss a simple model for repressor expression proposed by J. HASTY et al., 2000. The main actor in this model is the repressor molecule itself; we denote it by X. The advantage of the Hasty model is that it allows to quickly gain a qualitative insight, but it can be made fully quantitative as well. Here, we are mainly interested in the qualitative aspects, though.

A first simplification in the Hasty model is the assumption that one can neglect the operator binding site OR1. This is justified if one is only interested in the repressor kinetics: it means we do not want to model the full switch, but just the autoregulation mechanism of the λ-repressor.

In this case, the system can be described by six reactions. Four of them will be assumed as being *fast* when compared to the others; these are the association and dissociation of the transcription factors from their binding sites. They will be considered as equilibrium reactions, and are given by

- Repressor dimerization, $2X \leftrightarrow X_2$ with equilibrium constant K_1;

- Repressor dimer binding to the DNA promoter site OR2 on D, $D + X_2 \leftrightarrow DX_2$, with equilibrium constant K_2;

- Repressor dimer binding to the DNA promoter site OR3 on D, $D + X_2 \leftrightarrow DX_2^*$, with equilibrium constant K_3;

- Repressor dimer binding to the DNA promoter sites OR2 and OR3 on D, $DX_2 + X_2 \leftrightarrow DX_2X_2$, with equilibrium constant K_4.

The slow reactions in the system are transcription and degradation. Transcription is described by the irreversible reaction scheme

$$DX_2 + P \rightarrow_{k_t} DX_2 + P + nX \tag{7.1}$$

in which P is the DNA polymerase, and n is the number of repressor proteins per RNA transcript. The degradation reaction reads

$$X \rightarrow_{k_2} \phi. \tag{7.2}$$

We now define concentrations for all variables,

$$x \equiv [X], \, y \equiv [X_2], \, d \equiv [D], \, u \equiv [DX_2], \, v \equiv [DX_2^*], \, z \equiv [DX_2X_2]. \tag{7.3}$$

Considering the fast reactions as equilibrium reactions allows to write them as simple algebraic expressions,

$$y = K_1 x^2 \tag{7.4}$$
$$u = K_2 dy = K_1 K_2 dx^2 \tag{7.5}$$
$$v = \sigma_1 K_2 dy = \sigma_1 K_1 K_2 dx^2 \tag{7.6}$$
$$z = \sigma_2 K_2 uy = \sigma_2 (K_1 K_2)^2 dx^4. \tag{7.7}$$

It now remains to write down a rate equation for the repressor, which is given by

$$\dot{x} = \tilde{\alpha}u - k_d x + r.$$ (7.8)

Here, $\tilde{\alpha} = nk_t p_0$ is a constant containing k_t and the concentration of polymerase, p_0, which is assumed fixed. The term $-k_d x$ describes degradation of x while the last term, r, constitutes a basal transcription rate. In order to close the system of equations, a conservation law needs to be invoked. In fact, the total concentration of DNA promoter sites, d_T, is fixed.[1] Thus

$$d_T = d + u + v + z = d(1 + (1 + \sigma_1)K_1 K_2 x^2 + \sigma_2 K_1^2 K_2^2 x^4)$$ (7.9)

which leads to

$$d = \frac{d_T}{1 + (1 + \sigma_1)K_1 K_2 x^2 + \sigma_2 K_1^2 K_2^2 x^4}.$$ (7.10)

This use of expression Eq. (7.10) allows us to rewrite Eq. (7.8) in a succinct form containing only x as a variable,

$$\dot{x} = \frac{\alpha x^2}{1 + 2x^2 + 5x^4} - \gamma x + 1,$$ (7.11)

where repressor concentration and time have been rescaled to dimensionless quantities, and the values $\sigma_1 \approx 1$, $\sigma_2 \approx 5$, obtained from experimental estimates, have been used. The remaining parameters α and γ then constitute ratios of transcription rate and degradation rate relative to the basal transcription rate.

The mathematical discussion of this equation is extremely simplified by its one-dimensional character. In fact, we can rewrite it in the form

$$\dot{x} = -\partial_x V(x),$$ (7.12)

where $V(x)$ is the integral of the right-hand side of Eq. (7.12). $V(x)$ is now seen to serve as a 'potential energy' landscape, shown in Figure 7.2, a function which has two minima of different depth. The exact expression of $V(x)$ is

---

[1]This is true *in vivo*, for small concentrations, but even more so *in vitro*, for large concentrations.

given by

$$V(x) = V_0 - x + \frac{\gamma}{2}x^2 - \qquad (7.13)$$

$$\alpha \left[\frac{\left(\frac{1}{2} + \frac{i}{4}\right)\arctan\left(\frac{x}{\sqrt{\frac{1}{5} - \frac{2i}{5}}}\right)}{\sqrt{5 - 10i}} + \frac{\left(\frac{1}{2} - \frac{i}{4}\right)\arctan\left(\frac{x}{\sqrt{\frac{1}{5} + \frac{2i}{5}}}\right)}{\sqrt{5 + 10i}} \right]$$

Whichever is the lower of the two determines the steady state of the repressor system, i.e., either a state with a low concentration, or with a high concentration of repressor molecules.[2] Figures 7.2 and 7.3 are the main results of this subsection; it shows that repressor expression is a *bistable system* arising from the two competing mechanisms of repressor production and degradation.

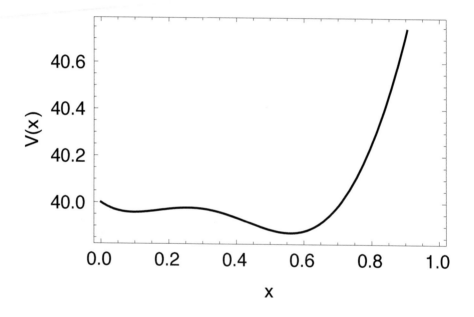

Figure 7.2 The effective potential of the λ-repressor, Eq. (7.13), for $\alpha = 50$ and $\gamma = 15$.

Hasty model II: looping included. We now include *looping* into the model. What this means is indicated in Figure 7.4. In the λ-phage a second, or left, operator region (OL) is found along the genome. This region, to which the corresponding genes are lacking, has a peculiar regulatory function (B. RÉVET

---

[2]This argument, as nice as it is, works indeed in general only for one variable. For more than one variable, a potential will only exist in special cases.

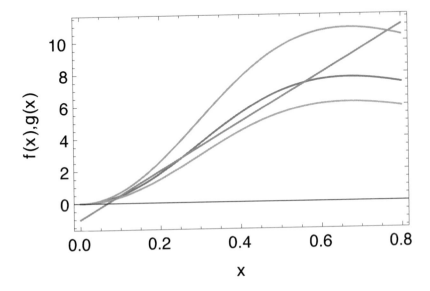

Figure 7.3 (**See color insert.**) Bistability in phage λ. Graph A shows the graphical solution of the stationary points of Eq. (7.12), i.e., $\dot{x} = 0$. In the figure, the three curves are the first term of the rhs of Eq. (7.12), denoted by $f(x)$, while the straight line is given by the function $g(x) = \gamma x - 1$, with $\gamma = 15$. The intersections of the graphs correspond to the extrema of $V(x)$, and can, e.g., be varied by changing γ. In the figure, the green curve is given by $\alpha = 70$, the blue curve by $\alpha = 50$, corresponding to Figure 7.2, while the yellow curve is for $\alpha = 40$.

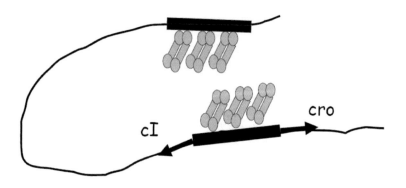

Figure 7.4 Looping in phage λ. (After M. PTASHNE, 2004.)

et al., 1999). Repressors that fix themselves at the OL-sites, are present to interact with their cousins sitting at OR-sites by forming a repressor octamer. Octamerization of repressors in solution is a very rare event, but the presence of the molecules on the DNA increases the formation rate enormously, since looping is a very effective mechanism of bringing different regions of DNA into contact (H. MERLITZ et al., 1998). It is therefore also a common regulatory mechanism in eukaryotes.

The Hasty model can be extended to account for DNA looping; one needs two additional assumptions. Again we simplify the structure of the left operator as we did for the right:

- the second operator unit equals the first: OR = OL; this in particular refers to all rate constants. OL is placed on a second DNA site, d_L.

- there is an additional kinetic process described by the complex formation between a doubly bound repressor at OR and the same at OL. This process has a rate K_5, and is assumed to be in equilibrium (fast compared to repressor production).

The presence of this additional mechanism modifies the nonlinearity in Eq. (7.12), which we now call $f_\ell(x)$, in the case looping. The change is due to the different possibilities of repressor fixation at either OR or at OL (d_R or d_L). Assuming the symmetric case with $d_R = d_L = d_\ell/2$ we find the conservation law

$$d_T = d_\ell(1 + (1 + \sigma_1)x^2 + \sigma_2 x^4) + \frac{\delta d_\ell^2}{4}x^8 \qquad (7.14)$$

which replaces Eq. (7.9), and where the factor δ derives from the looping process, i.e., the kinetic rate K_5. When the equation is solved for d_ℓ this yields, with $\sigma_1 = 2$ and $\sigma_2 = 5$,

$$d_\ell = 2 \left[\frac{1 + 2x^2 + 5x^4}{\delta x^8} \right] \left[\left[1 + \frac{d_T \delta x^8}{(1 + 2x^2 + 5x^4)^2} \right]^{1/2} - 1 \right]. \qquad (7.15)$$

Although expression (7.20) looks rather different from (7.10), there is in fact only a minor difference between them. This fact becomes obvious by expanding the square root in Eq. (7.20) for small x, which reduces (7.20) to the Eq. (7.10), while for large x the curves have the same asymptotics but a different amplitude.

This insensitivity to the presence of the nonlinear coupling is astonishing, but it allows to understand the role played by the looping-mediated coupling for the repressor dynamics. If one compares the stationary points of the repressor system with and without looping one sees that the effect of looping is simply to reduce the bistable region in which repressed and non-repressed transcriptional states coexist, see Figure 7.3. Given that the full operator region indeed has a significantly larger bistable region due to the presence of the operator site OR1 which we neglected in the calculation, the looping mechanism can be understood as an effective means to dynamically modulate the bistability of the repressor autoregulation.

As a further consequence, the looping mechanism also affects the fluctuations, which we have not discussed at all in our discussion. Since, with looping allowed for, the system can switch between the two states within a much narrower concentration range, the frequency of switching will consequently increase, while the amplitude diminishes. This is in accord with recent conclusions based on stochastic simulations (J. M. G. VILAR and S. LEIBLER, 2003).

It is worth noting that the simple model presented here neglects all spatial structure of the looping mechanism; the mechanism appears identical to that of an octameric complex binding at OR. But it is this process which is highly unlikely to occur since it has a very small equilibrium constant. By contrast, bringing dimer complexes at sufficiently distant sites into contact via looping occurs with a higher probability. The binding of repressor dimers at the OL region thus constitutes an effective mechanism to hold the molecules 'in stock' by placing them on the DNA, rather than letting them diffuse through solution.

7.2   DYNAMICS OF GENE REGULATORY NETWORKS

In this section we will look at models to describe the dynamics of gene regulatory networks, i.e., networks that via a specific wiring between their elements implement what has become known as the *'dogma of molecular biology'* as attributed to Francis Crick, namely

DNA makes RNA makes protein.

When we say 'DNA makes' we talk about a specific region on DNA which we call the gene; this is the DNA to be transcribed into mRNA. The latter is then translated into the protein product. J. PAULSSON (2005) has discussed the stochastic dynamics of such a process, invoking a population (n_1, n_2, n_3) of genes, mRNAs and proteins that follow simple growth and decay processes, except for the genes whose number is bounded above. In the following we will try to build a truly minimal description for this process involving first one gene only, and then combine several to form gene circuits. Thus we abstract from the full triad (gene - mRNA - protein) to a reduced model.

The gene gate model. We first define a very basic description of what a 'gene' can do. A minimal description for this process has been formulated by (Blossey, Cardelli and Phillips, 2006; 2008) in the gene gate model. A gene gate is an object which allows three types of actions:

- a gate which 'fires' without input, which corresponds to gene expression at a basal rate r;

- a gate which is blocked upon the binding a transcription factor;

- a gate which fires upon input, corresponding to the binding of a transcription factor, the activated case.

These cases are illustrated schematically in Figure 7.5.

This formal description makes it clear why the notion of a 'gate' is used: the analogy to computing by inputs and outputs is evident.

In this description, we have neglected the explicit distinction between mRNA and protein as input and output. This can be easily made; take this as a (*Task*). Let us now formalize the gene gate model in terms of reactions.

The action of gate a) can then be expressed as

$$G \to_\varepsilon G + P,$$ (7.16)

Figure 7.5 The three types of gene gates: a), left: constitutive transcription; b) middle: inhibition; c) right: activation of gene expression. In this graph, u and v denote the inputs and outputs of the gene gate.

where G is considered the "on"-state of the gene. The resulting proteins are degraded at a rate δ,

$$P \to_\delta 0. \tag{7.17}$$

The feedback of protein P on the gene can affect gene activity by putting the gene into its second state, G'. This is described by the reaction

$$G + P \to_r G' + P. \tag{7.18}$$

In the repressed case, the gene state G' is not productive and simply relaxes back to G at a rate η,

$$G' \to_\eta G. \tag{7.19}$$

G' is then thus the "off"-state of the gene. In the activated case (A), the gene is productive in state G' via

$$G' \to_\eta G + P \tag{7.20}$$

hence it relaxes back to gene state G while producing protein P, if the rate $\eta > \varepsilon$. Here, G and G' change roles: G is now the "off"-state (although never fully off) and G' is the "on"-state. We note that in the case $\eta < \epsilon$, the gene is actually repressed, but left with protein production at a finite rate η. This case does, in fact, not differ qualitatively from the case under (R).

This simple set of reactions with variables G, G', P and parameters $(\varepsilon, \delta, r, \eta)$ corresponds in terms of its complexity to a model introduced very early by PECCOUD and YCART (1995), which we abbreviate by PYM. The major difference between the gene gate model and the PYM is that the gene in the PYM *never* interacts with its proteins and reaction (7.18) occurs without the intervention of proteins. Further, the reaction (7.20) of case (A) does not occur. The main difference between the GGM and the PYM lies, in fact, precisely in reaction (7.20).

We will now formulate the master equation for the simplest gene circuit, the *self-regulated gene*. In the GGM, the reaction (7.20) leads in the master equation to state changes involving both the gene state and the protein number, which differs from the PYM.

Master equation for the self-regulated gene. The self-regulated gene is the gene circuit in which the gene product v and gene input u are identical: the output of the gene controls its input.

The master equation of the self-regulated gene has been formulated and solved by (Vandecan and Blossey, 2013), which we follow here. The master equations for the two versions of the GGM are readily written down. Denoting the basal state of the gene by 0, its second state by 1 and counting the protein number by n, one can introduce time-dependent probability distributions $p_{i,n}(t)$, with $i = 0, 1$. In case (R), (7.19) applies and the master equations read as

$$\forall n \geq 0 : \partial_t p_{1,n} = nr p_{0,n} - \eta p_{1,n} + \delta[(n+1)p_{1,n+1} - np_{1,n}]$$

$$\partial_t p_{0,0} = \eta p_{1,0} - \varepsilon p_{0,0} + \delta p_{0,1} \qquad (7.21)$$

$$\forall n \geq 1 : \partial_t p_{0,n} = \eta p_{1,n} + \varepsilon p_{0,n-1} - (\varepsilon + nr)p_{0,n}$$
$$+ \delta[(n+1)p_{0,n+1} - np_{0,n}]$$

while in case (A), for which Eq. (7.20) applies, we have

$$\partial_t p_{0,0} = -\varepsilon p_{0,0} + \delta p_{0,1}$$

$$\forall n \geq 1 : \partial_t p_{0,n} = -(\varepsilon + nr)p_{0,n} + \varepsilon p_{0,n-1} + \eta p_{1,n-1}$$
$$+ \delta[(n+1)p_{0,n+1} - np_{0,n}]$$

$$\forall n \geq 0 : \partial_t p_{1,n} = -\eta p_{1,n} + nr p_{0,n}$$
$$+ \delta[(n+1)p_{1,n+1} - np_{1,n}]. \qquad (7.22)$$

Introducing the generating functions

$$G_0(z, t) = \sum_{n=0}^{\infty} p_{0,n}(t) z^n \qquad (7.23)$$

and

$$G_1(z,t) = \sum_{n=0}^{\infty} p_{1,n}(t)z^n , \qquad (7.24)$$

the master equations above are transformed into first-order differential equations in t and z. For case (R), we have

$$\partial_t G_0 = \eta G_1 + \varepsilon(z-1)G_0 - (\delta(z-1) + rz)\partial_z G_0$$
$$\partial_t G_1 = -\eta G_1 - \delta(z-1)\partial_z G_1 + rz\partial_z G_0 \qquad (7.25)$$

while in case (A), we have

$$\partial_t G_0 = \eta z G_1 + \varepsilon(z-1)G_0 - (\delta(z-1) + rz)\partial_z G_0$$
$$\partial_t G_1 = -\eta G_1 - \delta(z-1)\partial_z G_1 + rz\partial_z G_0 . \qquad (7.26)$$

At this point it is instructive to compare the differential equations for the GGM to those of the PYM. The reaction (7.18), which describes the state change of the gene from G to G', is protein-number dependent and leads to a change from a term $\sim rG_0$ to a term $\sim z\partial_z G_0$; otherwise, the equations remain unchanged for case (R). In case (A), the notable difference is the presence of the term $\sim \eta z G_1$ in the equation for G_0, which results from reaction (7.20). Note that the corresponding η-dependent term in the equation for G_1 does not depend on z: the symmetry between the equations is broken.

We here solve the equations for case (A). Instead of solving for G_0 and G_1 separately we introduce the total generating function $G_T = G_0 + G_1$ with straightforward biological interpretation, allowing us to determine the mean value of the protein number and its fluctuations.

In case (A), the term $(z-1)$ can be placed in front in the equation for G_T, i.e.,

$$\partial_t G_T = (z-1)\Big(-\delta\partial_z G_T + \varepsilon(G_T - G_1) + \eta G_1\Big). \qquad (7.27)$$

One observes that a singularity arises at $z = 1$. The equality (7.27) is trivially satisfied at $z = 1$, because $G_T(1,t) = 1 \ \forall t$. A stationary solution requires $\partial_t G_T(z,t) = 0 \ \forall z, t$. For $z \neq 1$, this implies

$$-\delta\partial_z G_T + \varepsilon G_T + (\eta - \varepsilon)G_1 = 0 . \qquad (7.28)$$

Using Eq. (7.28), we can express G_1 and $\partial_z G_1$ in terms of $G_T(z,t)$. For the stationary solution, expression (7.26) can be rewritten in a second-order differential equation

$$\left(\delta(\delta - z(r+\delta))\right)\partial_z^2 g_T(z) + \left(-\delta\varepsilon + \delta z\varepsilon + rz\eta - \eta\delta\right)\partial_z g_T(z) + \eta\varepsilon g_T(z) = 0 \quad (7.29)$$

with $g_T(z) \equiv \lim_{t \to \infty} G_T(z,t)$. Substitution of $x = \delta(\delta - z(r+\delta))$ in Eq. (7.29) leads to the Kummer equation, familiar from the theory of ordinary differential equations,

$$x\frac{\partial^2}{\partial x^2}\tilde{g}_T(x) + (a + bx)\frac{\partial}{\partial x}\tilde{g}_T(x) + c\tilde{g}_T(x) = 0. \quad (7.30)$$

Here $a = (\varepsilon r + \delta\eta)/(r+\delta)^2$, $b = (\delta\varepsilon + r\eta)/(\delta^2(r+\delta)^2)$, $c = \eta\varepsilon/(\delta^2(r+\delta)^2)$ and $g_T(z) = \tilde{g}_T(x)$. Since $x = 0$ is a weak singular point of the differential equation (7.30), according to the method of Frobenius, a power series solution can be found around the singular point, i.e.,

$$\tilde{g}_T(x) = C_1 \sum_{n=0}^{\infty} a_n x^n + C_2 x^{1-a} \sum_{n=0}^{\infty} b_n x^n. \quad (7.31)$$

The substitution of the first term, i.e., the power series $\sum_{n=0}^{\infty} a_n x^n$, into Eq. (7.30), leads to the KummerM or the hypergeometric function $a_0 {}_1F_1(c/b, a; -bx)$. The second term corresponds to the KummerU function which can be discarded because the probability of n proteins does not tend to zero for $n \to \infty$. Consequently, the appropriate generating function $g_T(z)$ reads

$$g_T(z) = C_1' {}_1F_1(c/b, a; b\delta(z(r+\delta) - \delta)). \quad (7.32)$$

The coefficient C_1' is determined by the condition $\lim_{z\to 1} g_T(z) = g_T(1) = 1$ due to the continuity of the hypergeometric function ${}_1F_1$, so that $C_1' = {}_1F_1(c/b, a; b\delta r)$. The asymptotic solution in all parameters can be written as

$$g_T(z) = {}_1F_1\left(\frac{\eta\varepsilon}{\delta\varepsilon + r\eta}, \frac{\varepsilon r + \eta\delta}{(r+\delta)^2}; \frac{(\delta\varepsilon + r\eta)(z(r+\delta) - \delta))}{\delta(r+\delta)^2}\right) /$$
$$\qquad {}_1F_1\left(\frac{\eta\varepsilon}{\delta\varepsilon + r\eta}, \frac{\varepsilon r + \eta\delta}{(r+\delta)^2}; \frac{(\delta\varepsilon + r\eta)r)}{\delta(r+\delta)^2}\right). \quad (7.33)$$

From the total generating function (7.32) and Eq. (7.28), we can also separately derive the asymptotic solutions $g_0(z) = \lim_{t\to\infty} G_0(z,t)$ and $g_1(z) = \lim_{t\to\infty} G_1(z,t)$. The probability for measuring n proteins, $p_{0,n} + p_{1,n}$, is

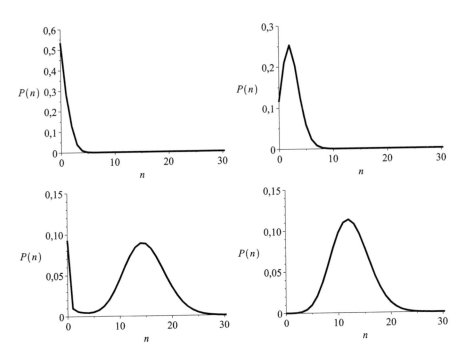

Figure 7.6 The probability distribution $P(n)$. Models and parameters from top left to bottom right: (TL), $\varepsilon = 0.01$; $\eta = 0.1$; $r = 1.0$; $\delta = 0.1$; (TR), $\varepsilon = 0.01$; $\eta = 0.1$; $r = 1.0$; $\delta = 0.04$; (BL), $\varepsilon = 15.0$; $\eta = 0.1$; $r = 0.001$; $\delta = 1.0$; (BR), $\varepsilon = 0.01$; $\eta = 0.1$; $r = 1.0$; $\delta = 0.008$. (Vandecan and Blossey, 2013). (Reprinted with permission from the American Physical Society.)

obtained from $(1/n!)\partial_z^n g_T(z = 0)$. We obtain an analytical expression for the stationary protein distribution $P(n)$

$$P(n) = \frac{1}{n!}\left(\frac{\delta\epsilon + r\eta}{\delta(r+\delta)}\right)^n \frac{\left(\frac{\eta\epsilon}{\delta\epsilon + r\eta}\right)_n}{\left(\frac{\eta r + \eta\delta}{(r+\delta)^2}\right)_n} \frac{{}_1F_1\left(\frac{\eta\epsilon}{\delta\epsilon + r\eta} + n, \frac{\epsilon r + \eta\delta}{(r+\delta)^2} + n; \frac{-\delta(\delta\epsilon + r\eta)}{\delta(r+\delta)^2}\right)}{{}_1F_1\left(\frac{\eta\epsilon}{\delta\epsilon + r\eta}, \frac{\epsilon r + \eta\delta}{(r+\delta)^2}; \frac{(\delta\epsilon + r\eta)r}{\delta(r+\delta)^2}\right)}$$

with $(.)_n$ being the *Pochhammer symbol*, $(x)_n = x(x+1) \cdot ... \cdot (x+n-1)$. The mean protein number E and its fluctuations V can also easily obtained from the first and second derivative of the generating function $g_T(z)$ with respect to z, evaluated at $z = 1$; we skip these results here.

Figure 7.6 shows the resulting probability distributions for both the repressed and the activated gene. We find two generic scenarios depending on relative parameter values. If degradation dominates over protein production, the protein probability is peaked at $n = 0$ (shown for (A), but likewise for (R)). In case (A), an increase of protein production lets a peak at finite n grow out of the peak at $n = 0$. In the repressed case, a bimodal appears with a maximum at $n = 0$ and a peak at finite n; the same behavior is found for the activated model with inversed rates, i.e., for $\varepsilon > \eta$, as shown in Figure 7.6 (bottom, left). For still larger values of protein production, the protein probability becomes monomodal with a peak at finite protein number in both cases (R) and (A). These analytical results of the gene gate models predict accurately the results of the stochastic dynamics simulations of these simple gates. For the parameters $r = 1$, $\epsilon = 0.01$, $\eta = 0.1$ and $\delta = 0.001$, the formula (7.34) reduces to a Poisson-type distribution,

$$P(n) \approx \frac{e^{-\lambda}\lambda^n}{n!} \tag{7.34}$$

with a mean $\lambda = (\delta\epsilon + r\eta)/(\delta(r + \delta)) \approx 99$ proteins in steady state, in accordance with simulations for the activated gene gate using the Gillespie algorithm (Blossey, Cardelli and Phillips, 2006).

Gene circuits from two and three genes. It is now clear how to pursue the construction of more complex gene gates: one needs to use the input of one gene as the input of another gene. Given the astonishing mathematical complexity the master equation of the self-regulated gene already displays, there are however only two ways to proceed: one either maps the problem onto a deterministic dynamics, or one simulates the master equation of the system with the Gillespie algorithm.

One can now go on to build more complex circuits from the basic gene gates, both deterministic and probabilistic. We will show the latter results for the *repressilator*. It is the most prominent example of a gene network, one which has been built artificially by (M. B. ELOWITZ and S. LEIBLER, 2000). The repressilator is a gene circuit which consists of *inhibitory gates*: the binding of a transcription factor at a binding site blocks the transcription of the corresponding gene. The 'wiring' of the repressilator is shown schematically in Figure 7.7.

Figure 7.8 shows a typical simulation run, provided suitable degradation rates are chosen. As was found in the experiments, the system can be made to oscillate, and all transcription factor concentrations go through clearly separated and synchronized rising and falling cycles (M. B. ELOWITZ and S. LEIBLER, 2000).

Deterministic dynamics of the gene gates. What if we want to discuss the deterministic dynamics of the gene gates? Let us follow (GIURANIUC, 2008). For each gene and protein, we introduce continuous variables, both for proteins p and for the gene states g and g', with the idea that we average over

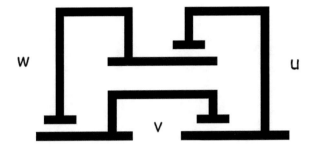

Figure 7.7 The wiring diagram of the repressilator.

an ensemble of cells. The self-regulated gene whose master equation we just discussed would then lead to two ordinary differential equations in the case of the self-inhibitory loop

$$\dot{p} = \varepsilon g - \delta p \qquad (7.35)$$

and

$$\dot{g} = \eta[1 - (1 + \nu p)g] \qquad (7.36)$$

where we used the relation $g + g' = 1$ to eliminate the state g'. Assuming further that we can also neglect the dynamics of the gene via $\dot{g} \approx 0$, we have $g = 1/(1 + \nu p)$ which, when plugged into the equation for p leads to

$$\dot{p} = \frac{\varepsilon}{1 + \nu p} - \delta p. \qquad (7.37)$$

This equation now presents us with a problem, as it only produces a single fixed-point for p. In order to remedy this deficit in the deterministic description, we need to consider multimerization of proteins, i.e. a nonlinearity in p in the denominator. We introduce this *Hill-exponent h* therefore by hand, and generalize to

$$\dot{p} = \frac{\varepsilon}{1 + \nu p^h} - \delta p. \qquad (7.38)$$

Can we now reproduce oscillations? Suppose we write down the three equations for the proteins a, b and c respecting the wiring of the repressilator circuit. We then have

$$\dot{a} = \frac{\varepsilon}{1 + \nu b^h} - \delta a. \qquad (7.39)$$

Figure 7.8 (**See color insert.**) Repressilator oscillations. Shown are the outputs over time.

and the two permutations $(a \to b, b \to c)$ and $(a \to c, b \to a)$. If we assume for simplicity that all parameters are identical, we find a symmetric fixed-point $a = b = c = \bar{a}$. Looking at the stability of this fixed point, we obtain the stability matrix *(Exercise: Verify!)*

$$\Gamma^{fp} = \begin{pmatrix} -\delta & -\kappa & 0 \\ 0 & -\delta & -\kappa \\ -\kappa & 0 & -\delta \end{pmatrix} \tag{7.40}$$

with

$$\kappa \equiv \frac{\varepsilon h \nu \bar{a}^{h-1}}{(1 + \nu \bar{a}^h)^2}. \tag{7.41}$$

The characteristic polynomial of Γ^{fp} is given by

$$(\delta + \lambda)^3 + \kappa^3 = 0 \tag{7.42}$$

which has one real root

$$\lambda_1 = -(\delta + \kappa) \tag{7.43}$$

and the other two complex roots are

$$\lambda_{2,3} = -\delta + (1 \pm i\sqrt{3})\frac{\kappa}{2}. \tag{7.44}$$

The condition for a *Hopf bifurcation* are thus given by

$$\delta = \frac{\kappa}{2} \tag{7.45}$$

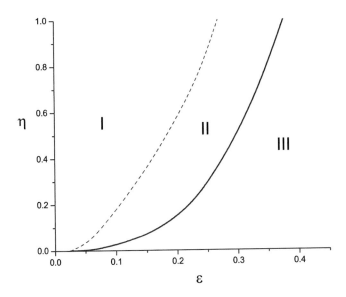

Figure 7.9 Parameter regimes for the deterministic repressilator dynamics. I, stable fixed point; II, stable fixed point for the reduced system, limit cycle for the full system; III, limit cycle (R. BLOSSEY and V.G. Giuraniuc, 2008). (Reprinted with permission from the American Physical Society.)

when the two complex roots turn imaginary. Making use of the fixed-point condition one finds the relation

$$\bar{a} = \left(1 - \frac{2}{h}\right)\frac{\varepsilon}{\delta}, \tag{7.46}$$

and hence we see that the condition

$$h > 2 \tag{7.47}$$

needs to be fulfilled for oscillations, i.e. for a fully stable *limit cycle*.

What happens if we take into account the full equations, i.e., when we keep the genes as dynamic quantities? We can then do a similar analysis from which follows that the condition on the Hill coefficient is weakened: $h > 4/3$ is then enough for oscillations to occur. Figure 7.9 compares the parameter regimes for the reduced and the full model. Figure 7.10 gives a comparison between the stochastic and deterministic limit cycles.

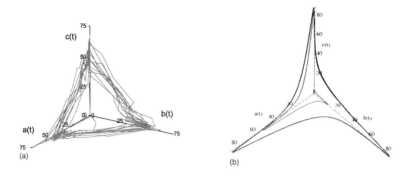

Figure 7.10 (**See color insert.**) (a) The limit cycle of the stochastic repressilator. Simulation parameters are $r = r_p = 1, \varepsilon = 0.1, \eta = 10^{-2}, \delta = 10^{-3}$. (b) The deterministic version for comparison with the reduced system in region III of Figure 7.9, parameters identical to the stochastic version, with Hill coefficient $h = 3$ (R. BLOSSEY and V.G. Giuraniuc, 2008). (Reprinted with permission from the American Physical Society.)

7.3 INTRINSIC NOISE

In this section we consider the dynamics associated with that of a binding site, say a promoter region for a transcription factor, present at a concentration c. We understand the binding and unbinding of the transcription factors as a process leading to fluctuations in the occupancy of the binding sites. How can we characterize these fluctuations?

We address this issue first in the framework of differential equations. The occupancy of the promoter site is, within a deterministic modeling setup, given by a kinetic equation for the time-dependent occupancy of the binding sites $n(t)$ for a given total concentration c of binding molecules. Following (W. BIALEK and S. SETAYESHGAR, 2005), we write

$$\dot{n} = k_+ c \left(1 - n(t)\right) - k_- n(t) . \tag{7.48}$$

At equilibrium, the binding is determined by detailed balance and associated with the binding free energy

$$F = k_B T \ln \left(\frac{k_+ c}{k_-}\right) . \tag{7.49}$$

We now assume that thermal fluctuations affect the rate constants. We linearize Eq. (7.48) around the mean value $\bar{n}$ to obtain

$$\dot{\delta n} = -(k_+c + k_-)\delta n + c(1-\bar{n})\delta k_+ - \bar{n}\delta k_-, \qquad (7.50)$$

and, in addition,

$$\frac{\delta k_+}{k_+} - \frac{\delta k_-}{k_-} = \frac{\delta F}{k_B T}. \qquad (7.51)$$

Together, this yields the expression

$$\dot{\delta n} = -(k_+c + k_-)\delta n + k_+c(1-\bar{n})\frac{\delta F}{k_B T}. \qquad (7.52)$$

We can Fourier transform the equation for $\delta n(t)$ and obtain the result (*Exercise!*)

$$\frac{\delta n(\omega)}{\delta F(\omega)} = \frac{1}{k_B T}\left[\frac{k_+c(1-\bar{n})}{-i\omega + (k_+c + k_-)}\right]. \qquad (7.53)$$

The total variance of the fluctuations can be obtained by invoking the *fluctuation-dissipation theorem*[3]

$$\langle(\delta n)^2\rangle = \int \frac{d\omega}{2\pi} S_n(\omega) \qquad (7.54)$$

where

$$S_n(\omega) = \frac{2k_+c(1-\bar{n})}{\omega^2 + (k_+c + k_-)^2}, \qquad (7.55)$$

which yields

$$\langle(\delta n)^2\rangle = \bar{n}(1-\bar{n}) = \frac{k_+c}{k_+c + k_c}(1-\bar{n}). \qquad (7.56)$$

Note that the inverse of the sum of the rates

$$\tau_c \equiv \frac{1}{k_+c + k_-} \qquad (7.57)$$

is the *correlation time* of the fluctuations.

---

[3] The fluctuation-dissipation theorem (FDT) in this context can be written as

$$\langle n(t)n(t')\rangle = \int \frac{d\omega}{2\pi} e^{-i\omega(t-t')} S_n(\omega)$$

.

We now want to generalize this result to the case where the concentration c of transcription factors is allowed to vary *spatially*, i.e., it obeys a diffusion equation with a sink term,

$$\partial_t c(\mathbf{x}, t) = D\Delta c(\mathbf{x}, t) - \delta(\mathbf{x} - \mathbf{x}_0)\dot{n}(t) \,. \tag{7.58}$$

One finds as a result the generalization of Eq. (7.53),

$$\frac{\delta n(\omega)}{\delta F(\omega)} = \frac{k_+ c(1 - \overline{n})}{k_B T} \left[\frac{1}{-i\omega(1 + \Sigma(\omega)) + \tau_c^{-1})} \right] \tag{7.59}$$

with

$$\Sigma(\omega) = k_+(1 - \overline{n}) \int \frac{d^3 k}{(2\pi)^3} \frac{1}{[-i\omega + Dk^2]} \,. \tag{7.60}$$

Task. Go through the steps of the above calculation. Discuss the properties of the Fourier-space integral; it is helpful to go back to the discussion of fluctuations in the Ising model in Chapter 1. Introduce a cutoff in k-space $\Lambda = \pi/a$.

The cutoff length a is a microscopic length which we identify with the size of the binding region. If we look for the low-frequency (long-time) limit of the function $\Sigma(\omega)$, we obtain

$$\Sigma(\omega \ll D/a^2) \approx \Sigma(0) = \frac{k_+(1 - \overline{n})}{2\pi Da} \,, \tag{7.61}$$

and we find for the *spectral density* of the occupancy fluctuations the result

$$S_n(\omega) \approx 2k_+ c(1 - \overline{n}) \frac{1 + \Sigma(0)}{\omega^2(1 + \Sigma(0))^2 + \tau_c^{-2}} \,. \tag{7.62}$$

The allowance for spatial fluctuations in the transcription factor concentration affects the correlation time, since now

$$\tau_c^{spatial} = \frac{1 - \overline{n}}{k_-} + \frac{\langle(\delta n)^2\rangle}{2\pi Dac} \,. \tag{7.63}$$

The diffusion contribution adds a minimum noise level given by the root-mean-square bound per time interval Δt

$$\delta n_{rms} > \frac{\overline{n}(1 - \overline{n})}{\pi Dac\Delta t} \,. \tag{7.64}$$

Let's now try to see whether this bound makes sense and plug in some numbers. In a bacterial cell, typically $N_{tf} \approx 100$ transcription factors are present in a cellular volume of $1\ \mu m^3$; their diffusion coefficient is $D \sim 3\ \mu m^2/s$. With the size of a promoter site of $a \approx 3\ nm$, one finds $\pi Dc \approx 3/s$. Thus the fluctuations in the site occupancy in a time interval Δt are given by

$$\frac{\delta n}{\overline{n}} > 0.55(1 - \overline{n}) \left(\frac{100}{N_{tf}} \right)^{1/2} \frac{s}{\Delta t} \,. \tag{7.65}$$

Experiments on *E. coli* show that about 10% efficiency of gene expression control can be achieved at small $\bar{n}$. In order to be consistent with the above limit, the promoter site must have an occupancy time on the order of one minute.

Stochastic cascades. In the previous section we studied how the interaction of a transcription factor with its binding site is affected by noise. In this section, we turn to the problem of how the noise that is affecting one component propagates in a network of several components. This occurs in transcriptional networks, i.e., the chain of biomolecules composed of transcription factors, DNA binding sites, mRNA products and proteins, as we discussed before. But this is equally true also in *signaling* or *metabolic networks*, where we speak about *pathways of biochemical reactions*, where at each level fluctuation effects can intervene. One might therefore simply expect that adding level upon level of noise on top of each other might always lead to a completely washed-out output signal. So the question arises what conditions must be satisfied such that a *graded input signal* can lead to an unambiguous, e.g., all or none, output signal?

We want to address this question for a simple model for a *signaling cascade* which was proposed by M. THATTAI and A. VAN OUDENAARDEN, 2002. While motivated by real biological cascades, such as phototransduction or protein kinase cascades, the model is not tuned to a direct comparison with such systems - which would require to build in much more detail - but rather to get a general idea about the noise robustness of reaction cascades. A recent application of these modeling ideas in a study of noise propagation in gene networks has been performed by J. M. PEDRAZA and A. VAN OUDENAARDEN, 2005.

The situation we have in mind is illustrated in Figure 7.11: The cascade is built up from a set of species of number Y_i, $i = 0, ..., N$ which are linked by differential amplification factors C_i which determine the response of Y_{i+1} to a change in Y_i. The input signal at Y_0 is read out at Y_N. At each level, the concentrations of the species are subject to a noise q_i. The model for this cascade that we will formulate is linear to begin with, and hence can be treated analytically using the methods of Chapter 5.

For the discussion, we need generic Langevin equations of the form

$$\dot{X} = g(X) + \eta(t) \tag{7.66}$$

where the noise $\eta(t)$ is defined by the moments

$$\langle \eta \rangle = 0, \quad \langle \eta(t)\eta(t+\tau) \rangle = q\delta(\tau). \tag{7.67}$$

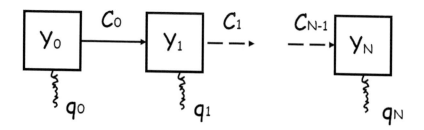

Figure 7.11 Generic model of a biochemical cascade. For the explanation of symbols, see text.

In these equations, we understand X the *number* of a chemical species in a cell. In the simplest case, we can assume that X is produced with a rate k_+ and destroyed with a rate k_-, where both processes are assumed to be Poisson-distributed. We then have

$$\dot{X}(t) = k_+ - k_- + \eta(t) \tag{7.68}$$

and we can compute the change in X in a time interval Δt as

$$\langle \delta X \rangle = N_+ - N_- = (k_+ - k_-)\Delta t. \tag{7.69}$$

Why did we assume production and destruction of X as Poisson processes? In order to make use of the relationship $\langle N_\pm^2 \rangle = \langle N_\pm \rangle$ which holds in this case. Thus we have

$$\langle \delta X^2 \rangle = (k_+ + k_-)\Delta t, \tag{7.70}$$

on the one hand, and, using Eq. (7.67) on the other we have

$$\langle \delta X^2 \rangle = \int_0^{\Delta t} \int_0^{\Delta t} dt\,dt'\, \langle \eta(t)\eta(t') \rangle = q\Delta t \tag{7.71}$$

such that we can identify

$$q = k_+ + k_- . \tag{7.72}$$

We now consider a protein product Y subject to a decay rate γ. We assume that it is produced proportional to an average number b per mRNA transcript, such that the equation reads as

$$\dot{Y} = kb - \gamma Y + \eta(t). \qquad (7.73)$$

In a steady-state, the variance is given by

$$\langle \delta Y^2 \rangle = (1 + b)\langle Y \rangle \qquad (7.74)$$

in which the factor b takes into account that the generally occurring 'bursts' in protein production will increase the variance above the Poisson level. From Eq. (7.73) we find after expansion $Y \to Y_0 + \delta Y$ for δX the equation for the fluctuations

$$\delta \dot{Y} + \gamma \delta Y = \eta(t). \qquad (7.75)$$

By Fourier transforming we obtain

$$(i\omega + \gamma)\delta\gamma(\omega) = \eta(\omega), \qquad (7.76)$$

and thus in time

$$\langle |\delta Y(\omega)|^2 \rangle = \frac{q}{\omega^2 + \gamma^2} \qquad (7.77)$$

and finally

$$\langle \delta Y^2 \rangle = \int_{-\infty}^{\infty} d\omega \frac{q}{\omega^2 + \gamma^2} = \frac{q}{2\gamma}. \qquad (7.78)$$

Since $\langle Y \rangle = kb/\gamma$, we obtain $q = 2kb(1 + b)$.

After these preliminaries we can finally go on to the cascade model. Its Langevin equation reads

$$\dot{Y}_i + \gamma Y_i = F_{i-1}(Y_{i-1}) \qquad (7.79)$$

where the decay rates are all assumed equal, and we simply put them equal to one. For the fluctuations we find

$$\delta \dot{Y}_i + \delta Y_i = C_{i-1}\delta Y_{i-1} + \eta_i \qquad (7.80)$$

where the noise variance is

$$\langle \eta_i(t)\eta_i(t + \tau) \rangle = q_i\delta(\tau). \qquad (7.81)$$

Employing the Fourier representation, we can write

$$\langle \delta Y_i^2(\omega) \rangle = \frac{q_i + C_{i-1}^2 \langle \delta Y_{i-1}^2(\omega) \rangle}{1 + \omega^2}. \qquad (7.82)$$

This equation is of the form

$$\zeta_i = \alpha_i + \beta_{i-1}\zeta_{i-1} \tag{7.83}$$

where

$$\alpha_i \equiv \frac{q_i}{1+\omega^2}, \quad \beta_i \equiv \frac{C_i}{1+\omega^2}, \quad \zeta_i \equiv \langle \delta Y_i^2(\omega) \rangle. \tag{7.84}$$

This recursion relation can be fully spelled out as

$$\zeta_N = \alpha_N + \beta_{N-1}\alpha_{N-1} + \beta_{N-1}\beta_{N-2}\alpha_{N-2} + \dots + \beta_{N-1}\cdots\beta_0\alpha_0. \tag{7.85}$$

Let us look for the upper bound of the variances in the limit of an infinite cascade, i.e., for $N \to \infty$. Introducing $q \equiv \max q_i$, $C \equiv \max |C_i|$, $\alpha \equiv q/(1+\omega^2)$ $\beta \equiv C/(1+\omega^2)$, we have

$$\zeta_N \leq \alpha(1 + \beta + \dots + \beta^{N-1}) + \beta\zeta_0 \tag{7.86}$$

from which the condition

$$\zeta_\infty \leq \frac{\alpha}{1-\beta}, \tag{7.87}$$

and consequently

$$\langle \delta Y_\infty^2 \rangle \leq \int \frac{d\omega}{2\pi} \frac{1}{1+\omega^2 - C^2} \leq \frac{q}{2\sqrt{1-C^2}} \tag{7.88}$$

follows.

The result of this calculation, Eq. (7.88), shows that the output fluctuations will stay bounded provided that $|C| \leq 1$, but they can be larger than the input fluctuations at any single cascade stage, due to the presence of the factor $1/\sqrt{1-C^2}$.

But, surprisingly, we will now see how this result can be used to attenuate the noise. For $|C| < 1$ the output noise will be independent of the input noise, if only the cascade is sufficiently long.

To demonstrate this, we again consider a cascade of species Y_i with low noise strength $q_i = q$ for all $i = 1, .., N$ which is subjected to a high noise at input, $q_0 > q$. The inverse Fourier transform of the recursion relation gives

$$\langle \delta Y_N^2 \rangle = \sum_{m=0}^{N} q_{N-m} \int \frac{d\omega}{2\pi} \frac{C^{2m}}{(1+\omega^2)^{m+1}} = \sum_{m=0}^{N} \frac{q_{N-m}}{2} \binom{2m}{m} \left(\frac{C}{2}\right)^{2m}. \tag{7.89}$$

The variance of the output signal at stage N of the cascade contains contributions from the cascade stage itself, and from the input carried along. With

Stirling's approximation for the factorial $m! \approx (m/e)^m \sqrt{2\pi m}$ this reads as

$$\langle \delta Y_N^2 \rangle = \frac{q}{2} \left(1 + \sum_{m=1}^{N-1} \frac{C^{2m}}{\sqrt{m\pi}} \right) + \frac{q_0}{2} \left(\frac{C^{2N}}{\sqrt{N\pi}} \right). \qquad (7.90)$$

While the first term in this expression increases with cascade length N, the second term is attenuated exponentially, since

$$\langle \delta Y_N^2 \rangle_{input} \sim \frac{e^{-\frac{N}{N_0}}}{\sqrt{N}} \qquad (7.91)$$

where

$$N_0 \equiv (-\ln(C^2))^{-1} \qquad (7.92)$$

is a sort of 'attenuation length scale'. It can serve as an estimate for the length of the cascade required to beat the noise on the input signal.

So much for a simple, linear model. The next step for us now is to see how fluctuation effects can work in systems which interact in a nonlinear fashion. We will discuss this for a, still fairly simple, feedback system in the next section.

Stochastic focusing. In this section, we discuss the stochastic dynamics of an intracellular negative feedback model system (J. PAULSSON and M. EHRENBERG, 2000). In this model, two species X and S are present that regulate each other's synthesis. The biological motivation is taken from bacterial genomes, the so-called *plasmids*. Plasmids are self-replicating genomes that are able to self-control their *copy number*. We want to see how a simple system of two components can achieve this.

The macroscopic dynamic equations describing the feedback system read

$$\dot{x} = \frac{kx}{1 + \alpha s} - x \qquad (7.93)$$

$$\dot{s} = k_s x - k_d s.$$

As before, the small letters x and s denote concentration variables. The interpretation of the equations is as follows: the X-molecules multiply autocatalytically (like the plasmids in the motivating example), while the S molecules inhibit X-synthesis by what is called *hyperbolic inhibition*. This is reflected in the factor

$$q \equiv \frac{1}{1 + \alpha s} \qquad (7.94)$$

in the equation for x: the presence of a high concentration of s inhibits the production of x.

Hyperbolic inhibition is a ubiquitous control mechanism arising in various reaction schemes (see J. PAULSSON et al., 2000). In the system given by Eq. (7.93), the parameters k and α set two characteristic concentration scales, and k_d determines how the steady state is approached: for small k_d, the approach is oscillatory, while for large values, s remains 'slaved' to x. Normalizing the equations with respect to the steady-state values via $x_r \equiv x/\bar{x}$ and $s_r \equiv s/\bar{s}$ leads to

$$\dot{x}_r = \frac{(k-1)(1-s_r)}{1+(k-1)s_r}x_r, \tag{7.95}$$

$$\dot{s}_r = k_d(x_r - s_r).$$

so that one immediately sees that in the limit $k \gg 1$ the rate of synthesis of X behaves has

$$\dot{x}_r \approx \left(\frac{1}{s_r} - 1\right)x_r. \tag{7.96}$$

Thus, high concentrations in S favour the decrease in X, and low concentrations in S favour the increase in X. The system cannot run away to high values of X, though, since the second equation couples the evolution of X directly to that of S.

We now want to look into what will happen to the fluctuations in X and S. For this we set up the *chemical master equation* which corresponds to the reactions between the S and X molecules. Suppose we start with m X-molecules and n S-molecules. The synthesis of X is governed by the rate $g_{mn} = km/(1+\alpha n)$. With the *step operator* $\mathcal{E}$ (see Part I, Chapter 2) one obtains the master equation in the form

$$\dot{p}_{mn} = (\mathcal{E}_m^{-1} - 1)g_{mn}\,p_{mn} + (\mathcal{E}_m - 1)m\,p_{mn} \tag{7.97}$$
$$+ \quad k_s m(\mathcal{E}_n^{-1} - 1)\,p_{mn} + k_d(\mathcal{E}_n - 1)n\,p_{mn} + p_{mn}\sum_{n=0}^{\infty} p_{1n}$$

for $m > 0$, $n \geq 0$; the last term arises from conditioning the distribution on $m > 0$. This choice reflects the fact that the state with $X = S = 0$ is an *absorbing state*. X can be considered a molecule whose absence signals cell death.

We now simplify this master equation by the additional assumption that the number of S molecules rapidly adjusts to the number of X molecules; it is thus considered a fast variable and can be eliminated from the equations. The dependence on n in Eq. (7.97) is thus dropped and the equation simplifies to

$$\dot{p}_m = (\mathcal{E}_m^{-1} - 1)g_m p_m + (\mathcal{E}_m - 1)m p_m + p_1 p_m . \tag{7.98}$$

This equation will now be analyzed for both noise-free and noisy signals, depending on the conditional variation of S.

When the conditional S-variation for a given value of X is negligible,

$$g_m = \frac{km}{1 + am(k_s/k_d)} . \tag{7.99}$$

If k is large (or, $am(k_s/k_d)$ small), g_m becomes constant for large m, and the number of X molecules remains Poisson distributed except at low averages $\langle m \rangle$. But when the conditional variation on S is not negligible - as a consequence of a noisy signal - Eq. (7.99) has to be replaced by

$$g_m = km \sum_{n=0}^{\infty} \frac{\bar{p}_{n|m}}{1 + an} . \tag{7.100}$$

The quasistationary conditional probabilities of n S-molecules given m X-molecules $\bar{p}_{n|m}$ are Poisson with average $\langle n \rangle_m = mk_s/k_d$, since all synthesis and degradation events are independent. The number of S-molecules at any given time represents the number of X-molecules in a probabilistic sense, and hence S is a 'noisy' slave to the slow variable X.

Figure 7.12 shows the variance of the X-molecules as a function of k for the two cases, assuming an average of $\langle m \rangle = 10$, and $\langle n \rangle_m = m$. One sees that for the second case, the variance in X can be reduced indefinitely, while it saturates for the former. This shows that it is indeed the variations of S that lead to a reduction in the variation of X. This effect has been dubbed

Figure 7.12 X-variance as a function of k for hyperbolic inhibition. [Reprinted with permission from J. PAULSSON and M. EHRENBERG; Copyright (2000) American Physical Society.]

stochastic focusing; it is the paradoxical effect that the fluctuations in one component effectively reduce those in another one.

Additional Notes

The lambda phage. The lambda phage is a superb model system for biologists, as well as a system to model for the biophysicist. The first modeling attempt, based on the free energies of binding of the transcription factors and, in particular, their cooperative behaviour, is by G. ACKERS et al., 1982. Stochastic analyses were performed by A. ARKIN et al., 1998 - but this is actually a topic for the next Chapter. The octamerization of the λ repressor by looping was found by B. RÉVET et al., 1999, see also the papers by I. B. DODD et al., 2001 and 2004. Finally, the paper by J. M. G. VILAR and S. LEIBLER, 2003, extends the work by Ackers and Arkin to account for the energetics and fluctuations of looping. Papers touching on further aspects of the λ switch are by E. AURELL and K. SNEPPEN, 2002; A. BAKK et al., 2004, and M. SANTILLÁN and M. C. MACKEY, 2004.

Deterministic modeling. The modelling of the dynamics of networks by nonlinear differential equations has a long history. Two of the pioneers are A. GOLDBETER and J. J. TYSON, see their work listed in the references.

Fluctuations, noise and robustness are becoming topics of increasing interest in the computational biology community. This is to a large part, but not exclusively, due to the influx of statistical physicists into the field. An early contribution is by H. H. MCADAMS and A. ARKIN (1997).

The effects of noise in the standard model of gene expression (DNA $\rightarrow$ RNA $\rightarrow$ protein) has been modelled and discussed with respect to experimental work in particular by J. PAULSSON (2004, 2005). Similar work is due to M. THATTAI and A. VAN OUDENAARDEN (2001) and by A. BECSKEI, B. B. KAUFMANN and A. VAN ODENAARDEN (2005).

Pi-calculus and process algebras. There is a whole literature developing on systems biology modelling based on approaches by theoretical computer scientists, in particular by specialists on programming languages. The seminal work is by A. REGEV (2002).

In order to model these gates we use a concept from theoretical computer science, a calculus called the π-*calculus* (R. MILNER (1999). We can understand the gate actions as 'communication' events: each gate can receive input or send outputs over *communication channels*. In our case, the messages sent over the channels are simple synchronizations between output and input actions at the two different gates. The ocurrence of an interaction (read: communication) requires both partners to have the corresponding input and output channels available.

In practice, this works as follows. The action of gate a) of Figure 7.5 which transcribes constitutively, will be written formally as

$$null(u) = \tau_c \,.\, (tr(u)|null(u)) \tag{7.101}$$

Note that this line should not be read as an equation but as a procedural description, as computer code. What is written on the right-hand side are the computational steps that are performed.

The notation means that the gate is considered a *process* with a parameter u which is its transcription product; there is no input. The notion of process in this context differs from the one commonly used in physics: this process is an object which can perform actions in time, like a running computer program. While the notation of the π-calculus may need some time of accommodation for a statistical physicist, the use of process algebras and calculi for modeling in systems biology is increasing. A big advantage of these approaches is their modularity (or, in technical language, *compositionality*). This means that the building blocks of the networks are easy to combine and modify, much easier than the case in the more classic approaches such as differential equations - an advantage that becomes crucial if one wants to model large systems.

Coming back to the formal description of the constitutive gate, the actions of this process are defined as follows:

- first perform a *stochastic delay* τ_c with a rate c. The rates can be determined according to the Gillespie algorithm simulating the underlying master equation of the stochastic dynamics, see Part I, Chapter 2;

- the separating dot means 'then';

- the term in brackets denotes two parallel processes x and y, $(x|y)$. The second process is a copy of the original process since processes are consumed after they have performed their actions;

- $tr(u)$ denotes the process which produces u, and has to be defined separately.

The process defining the transcription factor $tr(u)$ is written as

$$tr(u) = !u.tr(u) + \tau_\delta.0 \tag{7.102}$$

There are two actions now among which can be chosen: $+$ denotes choice in this context. The first process is an output action (represented by the exclamation mark '!'), followed by a copy of the original process, as already discussed before. The second action is a degradation process, denoted by 0, which occurs after a delay τ_δ, where δ again is a rate.

In the same style we can now introduce the regulated gates. The gate b) of Figure 7.5 which is blocked by receiving an input (a transcription factor) can be written as

$$neg(v, u) = ?v.\tau_\eta.neg(v, u) + \tau_c.(tr(u)|neg(v, u)) \qquad (7.103)$$

The second part of this expression should be clear: this is just the constitutive part. The first part means that the gate is ready to receive an input in the form of a transcription factor v which is followed by a stochastic delay; then the system returns to its initial state.

Finally the gate c) of Figure 7.5, whose expression activity is enhanced upon the action of a factor v is written as

$$neg(v, u) = ?v.\tau_\eta.(tr(u)|pos(v, u)) + \tau_c.(tr(u)|neg(v, u)) \qquad (7.104)$$

The symmetry between the first and the second parts of the statement is now evident; by comparison to gate b), gate c) both produces its product u constitutively, and by activation with factor v.

We now have all the elements to build the circuits. The wiring diagram of the repressilator is shown in Figure 7.7: three neg-gates are coupled. The wiring is expressed by the names of the products and hence the code describing the system is given by

$$neg(v, u)|neg(u, w)|neg(w, v) \qquad (7.105)$$

References

G.K. Ackers, A.D. Johnson, M.A. Shea
Quantitative model for gene regulation by λ phage repressor
Proc. Natl. Acad. Sci. USA **79**, 1129-1133 (1982)

A. Arkin, J. Ross, H.H. McAdams
Stochastic kinetic analysis of developmental pathway bifurcation in phage λ-infected Escherichia coli cells
Genetics **149**, 1633-1648 (1998)

E. Aurell, K. Sneppen
Epigenetics as a first exit problem
Phys. Rev. Lett. **88**, 048101 (2002)

A. Bakk, R. Metzler, K. Sneppen
Sensitivity of O_R in phage λ
Biophys. J. **86**, 58-66 (2004)

A. Becskei, B.B. Kaufmann, A. van Oudenaarden
Contributions of low molecule number and chromosomal positioning to stochastic gene expression
Nature Genetics **37**, 937-944 (2005)

W. Bialek, S. Setayeshgar
Physical limits to biochemical signalling
Proc. Natl. Acad. Sci. USA **102**, 10040-10045 (2005)

R. Blossey, L. Cardelli, A. Phillips
A compositional approach to the stochastic dynamics of gene networks
Transact. in Computational Systems Biology IV, 99-122 (2006)

R. Blossey, L. Cardelli, A. Phillips
Compositionality, stochasticity, and cooperativity in dynamic models of gene regulation
HFSP Journal **2**, 17-28 (2008)

R. Blossey, V.C. Giuraniuc
Mean-field versus stochastic models for transcriptional regulation
Phys. Rev E **78**, 031909 (2008)

I.B. Dodd, A.J. Perkins, D. Tsemitsidis, J.B. Egan
Octamerization of λ CI repressor is needed for effective repression of P_{RM} and efficient switching from lysogeny
Genes & Development **15**, 3013-3022 (2001)

I.B. Dodd, K.E. Shearwin, A.J. Perkins, T. Burr, A. Hochschild, J.B. Egan
Cooperativity in long-range gene regulation by the λ CI repressor
Genes & Development **18**, 344-354 (2004)

M. B. Elowitz, S. Leibler
Synthetic gene oscillatory network of transcriptional regulators
Nature **403**, 335-338 (2000)

C.P. Fall, E.S. Marland, J.M. Wagner, J.J. Tyson
Computational Cell Biology
Springer (2002)

A. Goldbeter
Biochemical Oscillations and Cellular Rhythms
Cambridge University Press (1996)

J. Hasty, J. Pradines, M. Dolnik and J.J. Collins
Noise-based switches and amplifiers for gene expression
Proc. Natl. Acad. Sci. USA **97**, 2075-2080 (2000)

H.H. McAdams, A. Arkin
Stochastic mechanisms in gene expression
Proc. Natl. Acad. Sci. **94**, 814-819 (1997)

H. Merlitz, K. Rippe, K. v. Klenin, J. Langowski
Looping dynamics of linear DNA molecules and the effect of DNA curvature: a study by Brownian dynamics simulation
Biophy. J. **74**, 773-779 (1998)

R. Milner
The π-Calculus
Cambridge University Press (1999)

J. Paulsson, O. Berg, M. Ehrenberg
Stochastic focusing: fluctuation-enhanced sensitivity of intracellular regulation
Proc. Nat. Acad. Sci. USA **97**, 7148-7153 (2000)

J. Paulsson, M. Ehrenberg
Random signal fluctuations can reduce random fluctuations in regulated components of chemical regulatory networks
Phys. Rev. Lett. **84**, 5447-4450 (2000)

J. Paulsson
Summing up the noise in gene networks
Nature **427**, 415-418 (2004)

J. Paulsson
Models of stochastic gene expression
Physics of Life Reviews **2**, 157-175 (2005)

J. Peccoud, B. Ycart
Markovian modeling of gene-product synthesis
Theor. Pop. Biol. **48**, 222-234 (1995)

J.M. Pedraza, A. van Oudenaarden
Noise propagation in Gene Networks
Nature **307**, 1965-1969 (2005)

M. Ptashne
A Genetic Switch. Third Edition: Phage Lambda Revisited
Cold Spring Harbor Laboratory Press (2004)

A. Regev
Computational Systems Biology: A Calculus for Biomolecular Knowledge
PhD-Thesis University of Tel Aviv (2002)

B. Révet, B. von Wilcken-Bergmann, H. Bessert, A. Barker, B. Müller-Hill
Four dimers of λ repressor bound to suitably spaced pairs of λ operators form octamers and DNA loops over large distances
Curr. Biology **9**, 151-154 (1999)

M. Santillán and M.C. Mackey
Why the lysogenic state of phage λ is so stable: a mathematical modeling approach
Biophys. J. **86**, 75-84 (2004)

M. Thattai, A. van Oudenaarden
Intrinsic noise in gene regulatory networks
Proc. Natl. Acad. Sci. USA **98**, 8614-8619 (2001)

M. Thattai, A. van Oudenaarden
Attenuation of noise in ultrasensitive signaling cascades
Biophys. J. **82**, 2943-2950 (2002)

J.J. Tyson, K. Chen, B. Novak
Network dynamics and cell physiology
Nature Rev. Mol. Cell Biol. **2**, 908-916 (2001)

Y. Vandecan, R.Blossey
Self-regulated gene: an exact solution for the gene gate model
Phys. Rev. E **87**, 042705 (2013)

J.M.G. Vilar, S. Leibler
DNA looping and physical constraints on transcription regulation
J. Mol. Biol. **331**, 981-89 (2003)

CHAPTER 8

Biological Networks: Space

8.1 EXTRINSIC VS. INTRINSIC NOISE

So far we have dealt with the interplay of noise in systems of few different, or many alike components. We now take a further step to distinguish more precisely the different contributions to noise, and how they can be disentangled in experiment. In particular, we add one important component to the discussion: space (which we have once considered before).

P. SWAIN et al., 2002 have introduced an explicit distinction between *extrinsic* and *intrinsic* sources of noise in the context of biological systems. We have seen already a definition of these notions from the point of stochastic dynamic systems - a clarification is thus needed of the relation between the two concepts with the same name. Let us discuss what Swain et al. refer to with these notions.

The difference between intrinsic and extrinsic noise according to the definition by Swain et al. is easy to grasp from the following example. Consider one particular gene and its protein product, let's call it A in a population of genetically identical cells. We want to relate this to the time evolution of the protein concentration or number in a single cell.

Even if all cells were identical at a given time, the molecular processes involved in the production of the protein (transcription and translation) are not identical in the cells; the production process of protein A will thus contain fluctuations that are intrinsic to the gene of protein A.

At the same time, the actors intervening in the production process, like the RNA-polymerase, are themselves gene products and also display cell-to-cell and time-dependent variations. Since their properties are not controlled by the production of protein A, one can consider their effect as being an extrinsic source of noise to the production of protein A. This means that the variations of the extrinsic factors (like the fluctuations in the number of polymerases) arise independently from the intrinsic variations, but obviously influence them. Extrinsic sources of noise in this sense are abundant in a cell, in which many processes are coupled.

The definition by Swain et al. is thus quite different from the one by van Kampen. In van Kampen's interpretation, systems like the ones from biology we discuss here have only *intrinsic noise* to begin with, since they are built up from molecular constituents; nowhere are deterministic processes in sight. The extrinsic noise by Swain et al. thus expresses the fact that the system can be separated by way of analysis into several noise-dependent mechanisms.[1]

Let us see where this assumption leads us.

We now formulate the mathematical description for the distinction between the two types of noise suggested by P. SWAIN et al., 2002. We represent the intrinsic and extrinsic properties by two vectors $\mathbf{I}$ and $\mathbf{E}$, whose components list the different noise sources. Further, let us identify by P_k the expression level of a specific gene in cell k.

From an experimental measurement taken on an ensemble of N cells, the P_k can be deduced and averaged to obtain the moments of the corresponding protein distribution in the form

$$\frac{1}{N} \sum_{k=1}^{N} P_k^m \approx \int d\mathbf{E}\, d\mathbf{I}\, P^m(\mathbf{E}, \mathbf{I}) p(\mathbf{E}, \mathbf{I}) \tag{8.1}$$

where p is the joint probability distribution for intrinsic and extrinsic variables, and $P^m(\mathbf{E}, \mathbf{I})$ is the measured expression level for particular values of $\mathbf{E}$ and $\mathbf{I}$.

Using the product rule of probability distributions we can separate this into contributions of intrinsic and extrinsic variables by invoking conditional

---

[1] The doubling of the notions of extrinsic and intrinsic with different meanings, once in the context of general stochastic processes and once in a specific biological context, is unfortunate. In the biological context, a better notion would probably have been to talk about *cis-noise* and *trans-noise*, where cis-noise is stochasticity in the same stochastic reaction pathway, and trans-noise is that exerted from one stochastic reaction pathway onto another. Both cis- and trans-noise are intrinsic in character.

probabilities

$$\frac{1}{N}\sum_{k=1}^{N} P_k^m \approx \int d\mathbf{E}\, p(\mathbf{E}) \int d\mathbf{I}\, P^m(\mathbf{E}, \mathbf{I})\, p(\mathbf{I}|\mathbf{E})\,, \qquad (8.2)$$

and we can introduce the definition of the average over intrinsic variables from the second integral

$$\langle P^m(\mathbf{E})\rangle \equiv \int d\mathbf{I}\, P^m(\mathbf{E}, \mathbf{I})\, p(\mathbf{I}|\mathbf{E})\,. \qquad (8.3)$$

By contrast, averages over extrinsic variables can be indicated with an overline in the form

$$\overline{\langle P^m\rangle} = \frac{1}{N}\sum_{k=1}^{N} P_k^m\,. \qquad (8.4)$$

which amounts to an average over both intrinsic and extrinsic noise sources.

If we want to quantify the noise strength by standard deviations over mean, we write for the total noise, as measured in experiment

$$\eta_{tot}^2 = \frac{\frac{1}{N}\sum_k P_k^2 - \left(\frac{1}{N}\sum_k P_k\right)^2}{\left(\frac{1}{N}\sum_k P_k\right)^2}\,, \qquad (8.5)$$

which is equivalent to

$$\eta_{tot}^2 = \frac{\overline{\langle P^2\rangle} - \left(\overline{\langle P\rangle}\right)^2}{\left(\overline{\langle P\rangle}\right)^2}\,. \qquad (8.6)$$

This can be rewritten as

$$\eta_{tot}^2 = \frac{\overline{\langle P^2\rangle - \langle P\rangle^2}}{\left(\overline{\langle P\rangle}\right)^2} + \frac{\overline{\langle P\rangle^2} - \left(\overline{\langle P\rangle}\right)^2}{\left(\overline{\langle P\rangle}\right)^2} = \eta_{int}^2 + \eta_{ext}^2\,. \qquad (8.7)$$

In order to distinguish experimentally between these two different contributions, one has to measure $\overline{\langle P\rangle^2}$. This quantity can be obtained as follows. If two identical copies of a particular gene were present in the same cell k, their expression levels were $P_k^{(1)}$ and $P_k^{(2)}$. These would have the same extrinsic sources of noise, but different intrinsic variables. Consequently,

$$\frac{1}{N}\sum_{k=1}^{N} P_k^{(1)} P_k^{(2)} \approx \int d\mathbf{E} \int d\mathbf{I_1} d\mathbf{I_2} P(\mathbf{E}, \mathbf{I_1}) P(\mathbf{E}, \mathbf{I_2}) p(\mathbf{I_1}, \mathbf{I_2}, \mathbf{E})$$

$$= \int d\mathbf{E}\, p(\mathbf{E}) \left[\int d\mathbf{I}\, P(\mathbf{E}, \mathbf{I}) p(\mathbf{I}|\mathbf{E})\right]^2$$

$$= \overline{\langle P\rangle^2}\,. \qquad (8.8)$$

The verification of this scenario can be made by fluorescent tagging of two genes controlled by identical regulatory sequences. These experiments have been performed by M. B. ELOWITZ ET AL., 2002, in *E. coli*. In the absence of intrinsic noise, the concentrations of both proteins, as measured by their fluorescence, would fluctuate in a correlated way. In a population of cells this then gives rise to a certain distribution of protein levels due to extrinsic noise alone. In the presence of intrinsic noise, the correlations between the two protein concentrations in each single cell decrease, reflecting different expression levels for the proteins.

Figure 8.1 shows a qualitative sketch of the experimental results in the form of a plot of noise levels as a function of a measure for population-averaged transcription rates, distinguishing between extrinsic and intrinsic noise contributions.

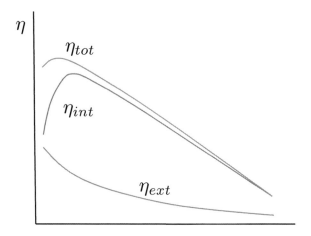

Figure 8.1 **(See color insert.)** *Cellular genetic noise* in *E. coli*. Qualitative behaviour of total, intrinsic and extrinsic noise as a function of fluorescence levels (population means) characterizing transcription rates.

8.2   THE TURING INSIGHT

However, a partial differential equation will be obtained which is thought to give a good approximation to the behaviour of certain kinds of chemical system. The differential equation has a number of parameters and it is necessary to find values for these parameters which will make the differential equation behave properly.

Alan Turing, The Development of the Daisy (1952)

In the 1950s the mathematician ALAN TURING hypothesized that the formation of biological structures may be driven by *chemical instabilities*. He proposed that the patterning results from an instability involving the reaction and diffusion of particular molecules, the *morphogens*. The idea that structure formation is based on the spatial distribution of chemical compounds and their interactions has proved to be a very profound one; meanwhile, several different variations of this scheme exist. We will go through the most important ones.

Morphogen gradients. The simplest way to generate a pattern is to assume that a morphogen is present at some localized source, and then diffuses into the surrounding tissue. The simplest chemical reaction that can happen is the decay of the morphogen due to degradation. Such a system is described by the differential equation

$$\partial_t \phi = D\partial_x^2 \phi - \mu\phi^\alpha + \eta\delta(x)\,, \qquad (8.9)$$

where ϕ is the concentration of the morphogen, D is its diffusion coefficient in the tissue, and μ the degradation coefficient. The delta-function term describes the localized source of the morphogen. Note that we have restricted ourselves to a one-dimensional case.

The parameter α allows to tune the degradation process: for $\alpha = 1$ one has the 'simple' degradation process while the value $\alpha = 2$ corresponds to an enhanced autodegradation. This case has recently been studied by (J. L. ENGLAND and J. CARDY, 2005). Eq. (8.9) then has a steady-state solution $\phi_0(x)$ given by

$$\phi_0(x) = \frac{6D}{\mu}\frac{1}{(x+\epsilon)^2} \qquad (8.10)$$

where $\epsilon \equiv (12D^2/\mu\eta)^{1/3}$. This solution corresponds to a gradient profile which decays algebraically outwards from the source.

A morphogen gradient can serve as a very simple means to structure space. Any mechanism capable of detecting *absolute* concentrations, say a value ϕ^*, naturally divides space into regimes $\phi > \phi^*$ and $\phi < \phi^*$. Although this mechanism seems astonishingly simple, it is actually realized in many systems (see, e.g., J. B. GURDON and P. Y. BOURILLOT, 2001).

Activators and inhibitors. More complex structuring effects can arise by considering not only one substance which decays, but also the interaction of two substances, one of which we call 'activator' a, while the second is called an 'inhibitor', h. A possible set of *reaction-diffusion equations* for such a system is given by

$$\partial_t a = D_a \partial_x^2 a - \mu a + \frac{\varrho a^2}{h} + \varrho_a , \tag{8.11}$$

$$\partial_t h = D_h \partial_x^2 h - \nu h + \varrho a^2 + \varrho_h , \tag{8.12}$$

where the terms $\propto \varrho a^2$ describe the auto- and crosscatalysis of the activator - as opposed to an autodegradation we had in Eq. (8.9) - while the term $\propto 1/h$ covers the action of the inhibitor; the remaining terms are degradation and source terms, as one verifies upon comparison with Eq. (8.9).

Equations (8.11), (8.12) are examples of a large class of models which were shown to generate pattern by a combination of *short-range activation* due to the catalytic nonlinearity $\propto a^2$ and a *long-range inhibition*. Several examples with different types of nonlinearities are discussed in the seminal paper by (A. GIERER and H. MEINHARDT, 1972). We will make a comment on the choice of nonlinearities later on.

We now turn to the technical discussion of the linear instability in the reaction-diffusion systems.

The Turing model. Turing was the first to see that *instabilities* in reaction-diffusion systems can lead to a pattern formation process. Let us discuss the appearance of an instability in the reaction-diffusion equations.

Suppose there are two chemical substances of concentrations ϕ_1 and ϕ_2, each of which is allowed to react according to a chemical reaction scheme (see Chapter 5). As before, the substances may diffuse. We now write down a general system for the reaction-diffusion equations

$$\begin{aligned} \partial_t \phi_1 &= f(\phi_1, \phi_2) + D_1 \Delta \phi_1 \\ \partial_t \phi_2 &= g(\phi_1, \phi_2) + D_2 \Delta \phi_2 \end{aligned} \tag{8.13}$$

wherein f and g are nonlinear functions of ϕ_1 and ϕ_2, $D_{1,2}$ are diffusion constants, and $\Delta \equiv (\partial_x^2 + \partial_y^2 + \partial_z^2)$ is the Laplace operator in Cartesian coordinates in three dimensions, generalizing what we wrote down before in eqs.(8.11), (8.12).

Suppose now that this system of equations has a *stationary point*, $\partial_t \phi_1 = \partial_t \phi_2 = 0$ at $(\phi_1, \phi_2) = (0,0)$. A linearization of eqs.(8.13) around this state in the form $\phi = \phi_0 + \delta\phi$ then leads to a linear system in the fluctuations $\delta\phi$; we drop the δ in the following to simplify notation:

$$\frac{d}{dt}\begin{pmatrix} \phi_1 \\ \phi_2 \end{pmatrix} = \begin{pmatrix} a_{11} & a_{12} \\ a_{21} & a_{22} \end{pmatrix} \cdot \begin{pmatrix} \phi_1 \\ \phi_2 \end{pmatrix} + \begin{pmatrix} D_1\Delta\phi_1 \\ D_2\Delta\phi_2 \end{pmatrix}. \quad (8.14)$$

The coefficients a_{ij} will in general depend on the value of ϕ_0. Now suppose that the system is confined to a cubic volume. Assuming that there is no flux of ϕ_1 or ϕ_2 entering the cube of side length ℓ, a *Neumann boundary condition* holds, i.e., the first spatial derivative of $\phi_{1,2}$ with respect to the wall normal vanishes.

The solutions of eqs. (8.14) can then be written in a general form as

$$\phi_1(\mathbf{x},t) = \sum_{n_1,n_2,n_3 \geq 0} c_{n_1,\dots,n_k}(t) \cos\left(\frac{2\pi}{\ell}x_1\right) \cdots \cos\left(\frac{2\pi}{\ell}x_3\right)$$

$$\phi_2(\mathbf{x},t) = \sum_{n_1,n_2,n_3 \geq 0} d_{n_1,\dots,n_k}(t) \cos\left(\frac{2\pi}{\ell}x_1\right) \cdots \cos\left(\frac{2\pi}{\ell}x_3\right)$$

$$(8.15)$$

where $c_{..}, d_{..}$ are the *Fourier coefficients* of the solution to eqs.(8.14). The terms in the sum are the *eigenmode solutions*, indexed by the set $N \equiv (n_1, n_2, n_3)$. The introduction of these expressions into eqs.(8.14) then leads to an infinite system of ordinary differential equations for the Fourier coefficients, which are given by

$$\frac{d}{dt}\begin{pmatrix} c_N \\ d_N \end{pmatrix} = \begin{pmatrix} a_{11} - (\frac{2\pi}{\ell})^2 D_1|N|^2 & a_{12} \\ a_{21} & a_{22} - (\frac{2\pi}{\ell})^2 D_1|N|^2 \end{pmatrix} \cdot \begin{pmatrix} c_N \\ d_N \end{pmatrix}$$

$$(8.16)$$

where $|N|^2 = n_1^2 + n_2^2 + n_3^2$.

We denote, in what follows, the (2×2)-matrix in Eq. (8.16) by M_N. For each non-negative 3-tuple N the stability of the eigenmodes is determined by the eigenvalues of M_N, $\lambda_N^{\pm}$. These can be expressed in terms of the trace and the determinant of the matrix M_N,

$$\lambda_N^\pm = \frac{1}{2}\left(TrM_N \pm \sqrt{(TrM_N)^2 - 4DetM_N}\right). \qquad (8.17)$$

For every $N \geq (0,0,0)$ the real and imaginary parts of Eq. (8.17) are bounded from above; the value

$$\Lambda \equiv \max Re\ \lambda_N^\pm \qquad (8.18)$$

thus is the upper bound of the spectral abscissae of the set of matrices $\{M_N\}$ (in the positive quadrant).

We see already by inspection of the entries of the matrix M_N that the diffusion terms may indeed *destabilize* a stable state of the homogeneous system, which may lead to a *symmetry breaking* of the global behaviour of the solutions to the nonlinear equations. As it turns out, for the generation of patterns, the two diffusion constants must significantly differ from each other, which may not be evident in many biological situations: why should two proteins, say, that diffuse in the cytoplasm have significantly different diffusion constants? One way out, as we will see below, is that one diffusion process is confined to the cell membrane, while the other occurs in the cytoplasm.

Returning to the eigenvalue problem of the Turing system, there are two types of eigenvalues that indicate the linear instability of the homogeneous state: in case of a *Turing instability*, Λ is positive and is one of the eigenvalues of M_N; if there is an oscillatory instability, there exists a mode with $Re\ \lambda_N^\pm = \Lambda > 0$, but for which $Im\ \lambda_N^\pm \neq 0$.

The occurrence of a Turing instability is, however, not generally sufficient to generate a *Turing pattern*, i.e., a time-*independent* spatial structure. We do not follow this trace here, since it is not necessarily a stationary pattern that is of interest to us in the description of spatial structures in biological systems. Also, patterning may arise in systems in which the conditions built into, e.g., the Gierer-Meinhardt model, like the production of an activator, need not be the case. We will now see a particular example, the Min system, to which this applies.

The Min system. We will now study an example of a particular pattern forming process, which is actually even more: a structuring process in bacteria, namely the cell division mechanism in *E. coli.* which is a rod-shaped bacterium of 2-6 μm length and 1-1.5 μm diameter. It divides roughly every hour: after replication of the DNA, a cell splits in half to form two daughter cells. The question is: how does the cell 'know' where its middle is?

The mechanism for cell division is based on the action of three proteins, the *Min* family: MinC, MinD and MinE. These proteins oscillate from one end to the other within the cytoplasm, and, furthermore, they move between the cytoplasm and the cytoplasmic membrane. Their concentration profiles determine the site for the cell division to occur.

The cell division itself is brought about by the formation of a contractile polymeric ring, called Z-ring, at the site of division. This ring is made from a protein named FtsZ, which is a homologue of tubulin which plays a similar role in higher organisms. FtsZ forms just underneath the cytoplasmic membrane, but how it generates the force needed for cell devision is still unknown. The Min-system serves to confine the location of this ring to midcell. If this mechanism is perturbed, ring formation can occur in the vicinity of the cell poles, and can lead to the formation of minicells devoid of DNA.

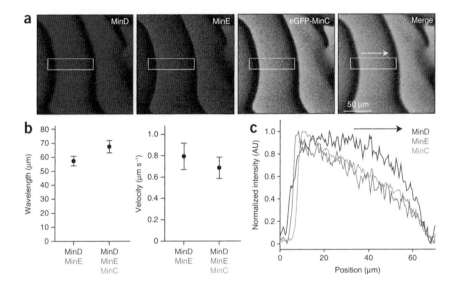

Figure 8.2 (**See color insert.**) MinD, MinE and eGFP-MinC in traveling waves in vitro. (a) Confocal fluorescence micrographs showing waves of MinD, MinE and MinC on a supported lipid membrane. (b) The influence of the presence of MinC on velocity and period of the protein waves. (c) Fluorescence intensity profiles of MinD, MinE and MinC acquired from the rectangular region shown in (a). Starting from the front of the wave (right), the density of MinC rises at a slope similar to that of MinE and also shows a similar sharp decrease at the rear of the wave. The detachment of MinC is shifted toward the front of the wave. [Reprinted with permission from Springer Nature from M. LOOSE et al. (2011).]

The action of the Min-system, as far as it is now known from molecular biology studies based on deletion mutants (i.e., bacteria in which corresponding genes have been knocked out), is as follows:

- MinC localized in the cytoplasmic membrane locally inhibits assembly of the contractile Z-ring. It remains in the cytoplasm and inactive in the absence of MinD;

- MinD binds MinC and recruits it to the cytoplasmic membrane;

- MinE drives MinD away from the membrane; the fact that it is driven mostly away from the bacterial midplane and hence allows for ring formation is a result of the protein dynamics.

Without MinE, the MinC/MinD couple would inhibit Z-ring formation everywhere, blocking cell division. In this case, long filamentous cells are observed. Without MinC, Z-ring formation cannot be inhibited anywhere, leading to inviable cells. Without MinD, neither MinC nor MinE are recruited to the cytoplasmic membrane and hence have less effect.

The wave dynamics of the Min-system has been visualized by fluorescence techniques, see Figure 8.2. Visualization has been possible inside the live bacteria (G. MEACCI and K. KRUSE, 2005), as well as in *in-vitro* systems, as shown in the figure.

In vivo, the MinC/Min D accumulate at a polar end of the bacterium on the cytoplasmic membrane. MinE forms a band at midcell which sweeps to the polar end and ejects the MinC/MinD into the cytoplasm. The ejected proteins then rebind at the other end of the bacterium. The MinE band, by contrast, dissociates at the pole and reforms at the center, and the whole process repeats towards the opposite cell pole. The observed oscillation period lasts 1-2 min, well below the cell division cycle length. The net effect of the dynamics is to minimize the concentration of MinC/MinD at midcell, and maximizing MinE concentration there, see Figure 8.3. The result shown there is obtained from a model developed by M. HOWARD ET AL., 2001, to which we will now turn.

The quantitative model for this dynamics based on the reaction-diffusion equation proposed by M. HOWARD et al., 2001, is based on four coupled equations for the densities of MinD and MinE in the cytoplasm and the cytoplasmic membrane. Since the MinC dynamics follows that of MinD, it does not need to be considered explicitly. The equations read (here, densities are ϱ_i with $i = e, E$ and $i = d, D$ for the membrane (small) and cytoplasm (capital)):

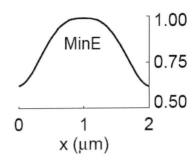

Figure 8.3 Concentration profiles of the Min proteins in E. coli. [Reprinted with permission from the American Physical Society (M. HOWARD et al., 2001).]

$$\partial_t \varrho_D = D_D \partial_x^2 \varrho_D - \frac{\sigma_1 \varrho_D}{1 + \sigma_1' \varrho_e} + \sigma_2 \varrho_e \varrho_d \tag{8.19}$$

$$\partial_t \varrho_d = \frac{\sigma_1 \varrho_D}{1 + \sigma_1' \varrho_e} - \sigma_2 \varrho_e \varrho_d \tag{8.20}$$

$$\partial_t \varrho_E = D_E \partial_x^2 \varrho_E - \sigma_3 \varrho_E \varrho_D + \frac{\sigma_4 \varrho_E}{1 + \sigma_4' \varrho_D} \tag{8.21}$$

$$\partial_t \varrho_e = \sigma_3 \varrho_E \varrho_D - \frac{\sigma_4 \varrho_e}{1 + \sigma_4' \varrho_D} \tag{8.22}$$

The first reaction terms in the equations for MinD describe the spontaneous association of MinD to the cytoplasmic membrane. Cytoplasmic MinD recruits cytoplasmic MinE to the membrane via the reaction term in the equation for MinE. Once there, MinE drives MinD into the cytoplasm, given by the second reaction term in the equations for MinD. The last term in the equations for MinE corresponds to the spontaneous dissociation from the membrane. All these terms have been modeled by introducing a parameter set $\{\sigma\}$, with parameter values given below.

This dynamics conserves protein number; there are no source or drain terms. The model exhibits a linear Turing-like *Hopf-instability*. Note that the mechanism of pattern formation in this system differs from the one proposed by TURING, and MEINHARDT and GIERER, since in this system there is no production of activators or an external influx of particles. The origin of the instability is the disparity between the diffusion rates between the cytoplasm and the cytoplasmic membrane; this has been idealized by setting the diffusion rates in the membrane to zero.

TABLE 8.1   Min system parameters

D_D $(\mu m^2/s)$	D_E $(\mu m^2/s)$	σ_1 (s^{-1})	σ_1' (μm)
0.28	0.6	20	0.028

σ_2 $(\mu m/s)$	σ_3 $(\mu m/s)$	σ_4 (s^{-1})	σ_4' (μm)
0.0063	0.04	0.8	0.027

Task. Analyze the linear stability of the system, testing for solutions of the form $e^{\lambda t + iqx}$. Take the parameter values used by M. HOWARD et al., 2001, listed in Table 8.1.

The equations (8.19) are valid only in a qualitative way, but establish a general basis for further improvement, which has occurred over the last years. But it should be clear that there is no general criterion to precisely decide on the form of the nonlinearities; instead, comparisons with different experiments need to be made in order to improve many details of the model.

Task. Discuss the difference between the models proposed by M. HOWARD et al., and H. MEINHARDT and P. A. J. DE BOER, both from (2001).

8.3   NETWORKS AS GRAPHS

After our study of biological systems in their real space we now look at them from an abstract perspective. The interactions between the different molecules that in one way or another regulate the functioning of the biological system at hand can be considered as an abstract space. In this way, *biological networks* of specific components can be defined and characterized. The starting point of our discussion of this abstract network structure is the random network, or *random graph* introduced by ERDÖS AND RENYI, 1959, which will serve us as a convenient *null model*. It is also called an *ER-graph*.

The ER-graph consists of n *nodes* or *vertices*, joined by *links* or *edges* between the vertices which are placed at random with independent probability p; an illustration is given in Figure 8.4. The construction allows to introduce the notion of an *ensemble of graphs* of n vertices, $\mathcal{G}(n, p)$, in which each graph is present with the probability corresponding to the number of its edges.

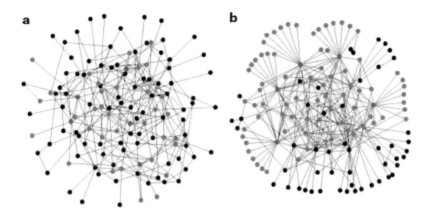

Figure 8.4 (**See color insert.**) Random network, or Erdös-Renyi graph a). The graph under b) represents a scale-free network, discussed in the text below. [Reprinted with permission from Springer Nature (R. ALBERT et al., 2000).]

The random graph has the following basic properties:

- The *degree* of a vertex is the number of edges connected to it. The *average degree* of a graph is then given by

$$z = 2 \times \frac{n(n-1)p}{2n} = (n-1)p \qquad (8.23)$$

where the factor of two takes into account that each edge is connected to *two* vertices. Hence there are no edges going out and ending at the same vertex.

- The probability p_k that a vertex in an Erdös-Renyi graph has degree k is given by the *binomial distribution*

$$p_k = \binom{n-1}{k} p^k (1-p)^{n-1-k} \qquad (8.24)$$

which in the limit $n \gg kz$ yields a *Poisson distribution*

$$p_k = \frac{z^k}{k!} e^{-z}. \qquad (8.25)$$

The degree distribution is thus strongly peaked around the mean z, and decays rapidly for large degrees.

Exercise. Show that ER-graphs are always *treelike*. A graph is called treelike if each of its *cycles* (or circuits, say internal loops) has at most one common vertex.

The question can be posed whether the random graph model applies to biological systems (or to other existing network-like systems, for that matter). An attempt in this direction was made by H. JEONG et al., 2000, in the context of a study of *metabolic networks* of 43 different organisms. A metabolic network is the network of biochemical compounds whose transformations are needed to sustain body function.

It was generally found that these large-scale networks display asymmetric degree distributions with rather *'fat'* tails, and could be characterized by algebraic behaviour for a range of degrees, k. In analogy to critical phenomena, think of the algebraic behaviour of correlation functions near a critical point discussed in Chapter 1, such networks have been dubbed *'scale-free'*.

In subsequent work, H. JEONG et al., 2001 considered the interaction network built from proteins in yeast *Saccharomyces cerevisiae*, consisting of 1870 edges and 2240 identified physical interactions among the proteins. Their finding is shown in Figure 8.5. Again, the degree distribution displays a fat tail, and the authors fit it to the expression

$$p(k) = \frac{A}{(k + k_D)^\gamma} \exp(-k/k_c) \tag{8.26}$$

where $k_D \approx 1$ and $k_c \approx 20$ are short- and large-scale cutoffs.

Evidently, the presence of cutoffs shows that the network is *not* strictly scale-free. This notion has indeed been much abused in the literature, and we will come back to it below in more detail. At present, it is important to also add that the algebraic behaviour is observed only over a quite limited range of k-values, sometimes hardly more than one order of magnitude. The exponent γ is quite typically found to lie in the range of $2 < \gamma < 3$.

Independent of the issue whether the graph is truly scale-free or not, already the appearance of a fat tail in the degree distribution obviously takes the protein interaction network out of the basic random graph paradigm. As a consequence, in order to understand these networks we need a new theory. Some additional observations provide further indication what this new theory should be able to deliver.

The *protein network* is an example of a network which is *undirectional*, i.e. the interaction of proteins does not distinguish between the partners. This does not hold for all networks, e.g., we may want to study a network of transcriptional interactions. In such a network, interactions are *directional*. A *transcription factor* will activate a specific gene (as we saw in Chapter 7), such that we need to have edges represent not just a basic 'interacts with', but the directedness from one node to another. A typical subgraph arising in such networks is shown schematically in Figure 8.6.

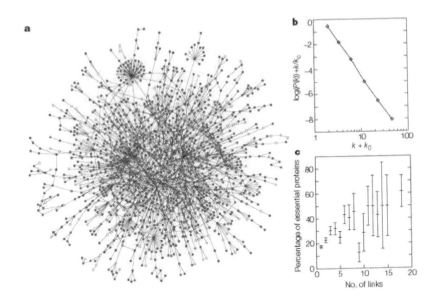

Figure 8.5 (**See color insert.**) A scale-free network of proteins in yeast. a) graphical representation of the network; b) evidence for the fat-tail distribution; c) essentiality of proteins as a function of the number of links. [Reprinted with permission from Springer Nature (H. JEONG et al. 2001).]

The fact that the construction of networks as graphs must reflect the particular identity of the interacting components explains immediately why the knowledge of degree distributions is not enough to characterize them. For directed networks, we have to distinguish between ingoing and outgoing degrees; but there are probably many deeper levels of structure as well. Also, it may not be sufficient to look at only one type of network (may be selected only because of the availability of data): S. MASLOV and K. SNEPPEN, 2002 have shown that the connectivity of protein interaction networks and transcriptional regulation networks in the yeast *Saccharomyces cerevisiae* show important correlations. The size of the networks used in this study amount to 4549 physical interactions (edges) between 3278 proteins (vertices) for the protein network, and 1289 positive or negative direct transcriptional interactions for 682 proteins.

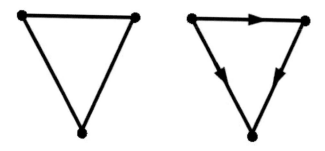

Figure 8.6 An undirected and a directed graph.

In order to go beyond the ER-graph-network paradigm, we now first list a number of important network characteristics and then show how to compute them from a given degree distribution by the method of generating functions. We then move on to a more general statistical mechanical theory of networks.

Probability generating functions. We have encountered generating functions already in previous chapters, so this section provides a novel application of this important tool.

We begin by taking the vertex degree distribution of the random network introduced before, p_k, and define, following (M.E.J. NEWMAN, 2003), the generating function

$$G_0(x) = \sum_{k=0}^{\infty} p_k x^k .$$
(8.27)

The vertex degree distribution p_k can be generated from G_0 by differentiation,

$$p_k = \frac{1}{k!} \frac{d^k G_0}{dx^k}\bigg|_{x=0} .$$
(8.28)

G_0 has the following properties:

1. If G_0 is properly normalized, then

$$G_0(1) = \sum_k p_k = 1.$$
(8.29)

2. The *average degree* $\langle k \rangle$ is found from computing

$$\frac{dG_0(x)}{dx}\bigg|_{x=1} \equiv G_0'(1) = \sum_k k p_k = \langle k \rangle .$$
(8.30)

3. For the n-th moment of p_k one has

$$\langle k^n \rangle = \sum k^n p_k = \left[\left(x \frac{d}{dx} \right)^n G_0(x) \right]_{x=1}. \tag{8.31}$$

We may of course also consider generating functions of probability distributions other than the degree distribution. For example, the probability that a vertex that can be reached along one edge from a starting vertex has $(k-1)$ outgoing edges is given by (*Exercise!*)

$$q_{k-1} = k \frac{p_k}{\sum_j j p_j}. \tag{8.32}$$

The average degree of the vertex then follows as

$$\sum_{k=0}^{\infty} k q_k = \frac{\langle k^2 \rangle - \langle k \rangle}{\langle k \rangle}. \tag{8.33}$$

This is the average degree of a vertex two edge distances away from the starting vertex. Based on this result, we can define the number of second neighbors in the graph by

$$z_2 = \left(\sum_k k q_k \right) \cdot \langle k \rangle = \langle k^2 \rangle - \langle k \rangle. \tag{8.34}$$

The probability generating function based on q_k finally reads as

$$G_1(x) = \sum_{k=0}^{\infty} q_k x^k = \frac{G_0'(x)}{z_1} \tag{8.35}$$

where $z_1 \equiv \langle k \rangle$.

We can now list further important network measures:

- The *clustering coefficient* C of a random graph is given by

$$C \equiv \frac{1}{nz} \left[\sum_k k q_k \right]^2 = \frac{z}{n} \left[c_v^2 + \frac{z-1}{z} \right]^2 \tag{8.36}$$

  where c_v is the *coefficient of variation* of the degree distribution (the ratio of standard deviation to mean).

- The *diameter d of a graph* is given by the maximum distance between any two connected vertices in the graph. It can be shown that the fraction of all possible graphs with n vertices and m edges for which $d \geq c \ln n$ for some constant c tends to zero in the limit $n \to \infty$.

- A *giant component* is a subgraph whose size scales linearly with the graph size. It corresponds to the *infinite cluster* in percolation theory.[2] The random graph has a phase transition (akin to the percolation transition on a lattice) in which a giant component is present whenever the *Molloy-Reed criterion* holds

$$\sum_{k=0}^{\infty} k(k-2)p_k = 0 \,. \tag{8.37}$$

- For networks which contain a giant component the *average vertex-vertex distance* ℓ is given by

$$\ell = \frac{\ln(n/z_1)}{\ln(z_2/z_1)} + 1 \,. \tag{8.38}$$

Task. Characterize correlations in the ER-graph.

Robustness. The *degree distribution* allows to characterize networks according to their *robustness* or resilience to attack. R. ALBERT et al. (2000) compared the properties of an ER-graph and a scale-free graph upon edge removal. These networks differ in a characteristic way: since vertices and edges in ER-graphs are more 'alike', their properties such as the average degree change fairly little when edges are removed randomly or even on purpose, i.e., when vertices with a high degree are removed preferentially, simulating an attack on the network. By contrast, the removal of few, highly connected nodes in a scale-free network affects its structure very much.

Exercise. Study the robustness of the network described by the generating function

$$G_0(x) = \frac{1}{n}\left((n-1)x^3 + x^n\right) \,. \tag{8.39}$$

Picture this network. How does its diameter change as a function of n?

We are now ready for a more systematic approach, along the lines of equilibrium statistical mechanics of Chapter 1. As there, a notion of entropy will prove to be highly useful.

---

[2]Percolation is a *geometric phase transition*: consider the case of a two-dimensional regular lattice with all sites occupied. Now one places bonds between the sites with a given probability. A question one can pose now is at what concentration of occupied sites will a cluster arise which spans the whole lattice: this is the infinite cluster. Its appearance (the percolation transition) and the statistical properties of this cluster share many features of thermal phase transitions, but there is of course no free energy in this case.

8.4   STATISTICAL MECHANICS OF NETWORKS

A statistical mechanical theory of networks can be developed for *exponential random graphs* (J. PARK and M.E.J. NEWMAN, 2004). If G is an element of a set of graphs $\mathcal{G}$, we can obtain its probability distribution $P(G)$ by maximizing the Gibbs entropy

$$S = -\sum_{G \in \mathcal{G}} P(G) \ln P(G) \tag{8.40}$$

under the constraints $\sum_G P(G) x_i(G) = \langle x_i \rangle$ and $\sum_G P(G) = 1$, where x_i, $i = 1, ..., r$ are a set of graph observables. The derivation of $P(G)$ can now be performed in complete analogy to our discussion of the thermal ensembles in Chapter 1 (*Task!*), leading to the result

$$P(G) = \frac{e^{-\sum_i \theta_i x_i(G)}}{Z} = \frac{e^{-H(G)}}{Z}. \tag{8.41}$$

As in the thermal case, the expression Eq. (8.41) involves a set of Lagrange multipliers θ_i.

Examples. The simplest example is the random graph with a fixed number of vertices n. If we want to characterize the graph only by the mean number of edges, $\langle m \rangle$, the Hamiltonian in Eq. (8.41) is chosen to be

$$H(G) = \theta m(G) \tag{8.42}$$

We can now evaluate the partition function Z for an ensemble of simple undirected graphs on n vertices. An $(n \times n)$ adjacency matrix can be defined by

$$\sigma_{ij} = \begin{cases} 1 & i \leftrightarrow j, \\ 0 & else \end{cases} \tag{8.43}$$

where the double-arrow symbol $\leftrightarrow$ denotes i connected to j. With the adjacency matrix we can represent the number of edges as

$$m = \sum_{i<j} \sigma_{ij}, \tag{8.44}$$

and hence compute the partition function Z

$$\begin{aligned} Z &= \sum_G e^{-H} = \sum_{\{\sigma_{ij}\}} \exp\left(-\theta \sum_{i<j} \sigma_{ij}\right) \\ &= \prod_{i<j} \sum_{\sigma_{ij}=0}^{1} e^{\theta \sigma_{ij}} = \prod_{i<j}(1+e^{-\theta}) = [1+e^{-\theta}]^{\binom{n}{2}} \end{aligned} \tag{8.45}$$

We can also define the analogue of the free energy by $F \equiv -\ln Z$ and find the result

$$F = - \binom{n}{2} \ln(1 + e^{-\theta}),$$
(8.46)

such that the expected number of edges in the graph is

$$< m > = \frac{1}{Z} \sum_G m e^{-H} = \frac{\partial F}{\partial \theta} = \binom{n}{2} \frac{1}{1 + e^{\theta}}.$$
(8.47)

In the last equation, we may want to redefine $p = (1 + e^{\theta})^{-1}$ to simplify notation. The probability of a graph in this ensemble is then given by

$$P(G) = \frac{e^{-H}}{Z} = p^m (1 - p)^{\binom{n}{2} - m}.$$
(8.48)

$P(G)$ is the probability for a graph in which each of the $\binom{n}{2}$ possible edges appears with probability p; this is just a Bernoulli random graph.

Further examples. We list some further examples; the computation of their properties is left as *Exercises*. (J. PARK and M.E.J. NEWMAN, 2004).

Specifying degrees. If we choose the vertex degrees k_i as observables,

$$H = \sum_i \theta_i k_i,$$
(8.49)

we can rewrite this using the adjacency matrix σ_{ij} as

$$H = \sum_{ij} \theta_i \sigma_{ij} = \sum_{i<j} (\theta_i + \theta_j) \sigma_{ij}.$$
(8.50)

The partition function is given by

$$Z = \prod_{i<j} (1 + e^{-(\theta_i + \theta_j)}).$$
(8.51)

Directed graphs. We now change the graph ensemble $\mathcal{G}$ to contain directed graphs; we have to adopt the adjacency matrix accordingly. It now contains an entry 1 if an edge exists in the graph from j to i ($j \to i$). For the choice $H = \theta m$ one has

$$Z = \prod_{i \neq j} \sum_{\sigma_{ij}=0}^{1} e^{-\theta \sigma_{ij}} = [1 + e^{-\theta}]^{2\binom{n}{2}}$$
(8.52)

Fixed edges. If one chooses for $\mathcal{G}$ the set of graphs with both fixed number of vertices n and edges m,

$$Z = \sum_G \delta(\tilde{m}, m(G))e^{-H} \tag{8.53}$$

where $\tilde{m}$ is the desired number of edges. This construction corresponds to a canonical ensemble (J. BERG and M. LÄSSIG, 2002).

Network growth. We have now learnt how to characterize statistically graphs or networks of more general nature than simple random graphs, so that we feel ready to come back to the problem of protein network structure. We can determine, e.g., degree distributions empirically based on data and characterize the networks by various measures. A detailed analysis by J.-D. HAN et al. (2005) demonstrated the pitfalls one can get into by an uncritical data analysis: the limits on the sampling range alone can already lead to wrong conclusions on the degree distributions.

Whatever we do with the existing incomplete data, this does not, clearly, tell us how they came about in the first place; it is hence of interest to model the evolution of a network. For the protein-protein interactions networks, e.g., known mechanisms of network evolution are *gene duplication* and *mutation*. In order to build a simple model for the dynamics of network growth, two basic assumptions can be made (R. ALBERT and A.-L. BARABÁSI, 2000; J. KIM et al., 2002):

- **Vertex duplication.** Vertices (i.e., new proteins) are added, one after the other. A new vertex duplicates a previously existing vertex which is chosen randomly, and links to its neighbors are placed with a probability $1 - \delta$.

- **Diversification.** Each new vertex links to any previous node with probability β/n, where n is the current total number of vertices in the network.

A protein interaction network generated from such mechanisms thus mimics the underlying evolutionary mechanisms. In these, mutations occur both at the duplication and diversification levels, if the parameters $\beta, \delta > 0$.

The average vertex degree of such a network G can be estimated as follows. In each growth step, the average number of edges $\langle m \rangle$ increases by $\beta + (1 - \delta)G$. Thus,

$$\langle m \rangle = (\beta + (1 - \delta)G)n. \tag{8.54}$$

Since generally the relation $G = 2\langle m \rangle/n$ holds, we find, eliminating $\langle m \rangle$

$$G = \frac{2\beta}{2\delta - 1}, \tag{8.55}$$

which can only hold for $\delta > \delta_c = 1/2$. Below this threshold, the number of links grows according to

$$\frac{d\langle m \rangle}{dn} = \beta + 2(1 - \delta)\frac{\langle m \rangle}{n} \tag{8.56}$$

which together with the relation $G(n) = 2\langle m \rangle(n)/n$ yields the scaling dependencies

$$G(n) = \begin{cases} \text{finite} & \delta > 1/2, \\ \beta \ln n & \delta = 1/2, \\ C \times n^{1-2\delta} & \delta < 1/2. \end{cases} \tag{8.57}$$

Without diversification ($\beta = 0$), a finite average vertex degree is therefore only found if $\delta > 1/2$, illustrating the important role mutations play.

The case of $\delta > 1/2$, $\beta > 0$ can be studied in more detail by the rate equation for the evolution of the number of vertices of degree k, $P_k(n)$ when the network as a whole has n vertices. The degree of a node increases by one at a rate $a_k = \beta + (1 - \delta)k$. We can then write down a rate equation of the form

$$\frac{dP_k(n)}{dn} = \frac{1}{n}(a_{k-1}P_{k-1} - a_k P_k) + G_k \tag{8.58}$$

where the first two terms account for the increase of a vertex degree by one. The last term is a source term for the introduction of new vertices with k edges, with a of the edges created by duplication and $b = k - a$ by diversification. The probability of the duplication process is

$$g_{dup} = \sum_{s \geq a} x_s \binom{s}{a}(1 - \delta)^a \delta^{s-a} \tag{8.59}$$

where $p_s = n_s/n$ is the probability of a vertex of degree s chosen for the duplication process. The probability of diversification is

$$g_{div} = \beta^b \frac{e^{-\beta}}{b!} \tag{8.60}$$

such that the full expression for G_k is given by

$$G_k = \sum_{a+b=k} \sum_{s=a} x_s \binom{s}{a}(1 - \delta)^a \delta^{s-a} \beta^b \frac{e^{-\beta}}{b!}. \tag{8.61}$$

Since the n_k grow linearly in n, we can plug the relation $P_k(n) = np_k$ into the rate equation and obtain

$$\left(k + \frac{1+\beta}{1-\delta}\right)p_k = \left(k - 1 + \frac{\beta}{1-\delta}\right)p_{k-1} + \frac{G_k}{1-\delta}, \tag{8.62}$$

which looks very much like a recursion relation for the p_k - but it is not, since G_k depends on all p_s for $s \geq k$. It can, however, be turned into a recursion relation in the limit $k \to \infty$. In this limit, the main contribution to G_k arises for small values of b, and the summand is sharply peaked around $s \approx k/(1-\delta)$. We may then replace the lower limit by $s = k$ and p_s by its value at $s = k/(1-\delta)$. Further, anticipating that p_k decays as $k^{-\gamma}$ we introduce the *ansatz* $p_s = (1-\delta)^\gamma p_k$ and finally simplify G_k according to

$$G_k \approx (1-\delta)^\gamma p_k \sum_{s=k} x_s \binom{s}{k} (1-\delta)^k \delta^{s-k} \sum_{b=0}^{\infty} \beta^b \frac{e^{-\beta}}{b!} = (1-\delta)^{\gamma-1} p_k \quad (8.63)$$

since the binomial sum equals $(1-\delta)^{-1}$.

Summing this all up, in the limit $k \to \infty$, p_k is found to have a power-law behaviour $p_k \sim k^{-\gamma}$ with the value of γ fixed by the relation

$$\gamma(\delta) = 1 + \frac{1}{1-\delta} - \frac{1}{(1-\delta)^{2-\gamma}} \quad (8.64)$$

We thus find that $\gamma(\delta)$ sensitively depends on δ, but *not* on β. Choosing a value of $\delta = 0.53$, as can be suggested from observations (A. WAGNER, 2001), a value of $\gamma = 2.373..$ is found, in accord with what we had indicated before.

A particularly important point to notice is that this result is an *asymptotic* one. The convergence to the asymptotic limit in which the power laws reign is very slow. Even for vertex numbers of $n = 10^6$, the power-law regime is only reached over two orders of magnitude of degrees k. Compare this with the data for real protein networks.

8.5 SMALL WORLDS

Another important network paradigm that has been uncovered in the last years is that of *small-world networks* (D. J. WATTS and S. H. STROGATZ, 1998). The authors studied a random rewiring process of a completely regular network. Initially, the model network is placed on a one-dimensional ring and up to k neighbouring points are linked. Then links are broken and rewired at random across the lattice. Upon the introduction of only a few of these 'shortcuts', a significant clustering property of the network arises, characterizing an intermediate stage between complete regularity on the one side, and complete randomness on the other.

The important feature of the small-world model is that distant points on the network get access to each other: only a small number of shortcuts will thus allow the network, although being still fairly sparse, to 'communicate' efficiently across all nodes. This phenomenon has acquired some fame through the famous story of 'six degrees of separation', referring to the number of acquaintances needed to pass between two randomly selected people in the world. M. E. J. NEWMAN at al., 2000, formulated an exactly solvable mean-field model for the small-world network, but despite being 'just a mean-field solution', it is already quite demanding.

Motifs and modules. Biological networks may display particular global features, but they will also be characterized by the properties of local neighbourhoods of vertices. This has been noticed via the occurrence of particular *network motifs* (R. MILO et al., 2002, S. SHEN-ORR et al., 2002).

Figure 8.7 shows a prominent example, the *feedforward loop*.[3] While in a random graph subgraphs are more likely to be trees, functional subgraphs of networks are more likely to have a higher number of edges than a subgraph in a random network. Again the distinction between directed and undirected graphs matters.

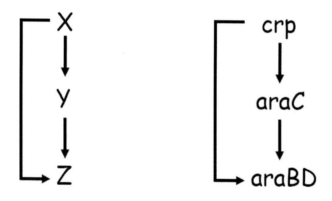

Figure 8.7 A feedforward loop. Left: basic structure of the loop; right: an example of a feedback loop in the gene network of *E. coli.*

There have been meanwhile a number of studies focussing on network motifs, e.g. for *E. coli* and *S. cerevisiae* (E. YEGER-LOTEM et al., 2004). In particular the feedforward loop has interesting features which have been recently discussed (S. MANGAN and U. ALON, 2003; S. KALIR et al., 2005). Eight

---

[3]Note that the innocently looking graph to the right in Fig. 8.7 has the same topology!

different types of feedforward loops have been identified, depending on the actions of the transcription factors (activating or repressing).

In *E. coli*, aside from the feedforward loop, two other motifs have been detected as being statistically overrepresented in the data: the so-called single input module, in which a single transcription factor regulates a set of transcriptional units, the *operons*, see our discussion of the phage lambda. A third motif is the *dense overlapping regulon*, in which a number of operons is regulated by a set of transcription factors.

A review of the different types of motifs that have been detected is by D. M. WOLF and A. A. ARKIN, 2003). The notion of a motif generalizes, on a higher network level, to that of a *network module*, comprising as separately functioning network elements (E. RAVASZ et al., 2002). The distinction between a motif and a module is not evident, as Wolf and Arkin note themselves.

RINs: residue interaction networks. The network of amino acids, or *residue interactions*, is an example of such a small world network. The motivation to consider the networks formed by amino acids has been to find out what residues could play pivotal roles for the function of the protein, in particular if the protein shows *allostery*, i.e. a conformational change in one region upon binding of ligand in another region of the protein.

The relevance of protein residues is typically characterized by two measures. The first is the *closeness centrality* (A. DEL SOL et al., 2006) which is given by

$$C_k = \frac{n-1}{\sum_{i \neq k} n(i,k)} \tag{8.65}$$

where n is the total number of residues and $n(i,k)$ is the shortest path distance between residues i and k. Residues are defined to be in contact via some distance measure in real space: typically a condition of $d < 5$ Åis employed, but this can vary upon context. The residues are then scored via the relation

$$z_k = \frac{C_k - \overline{C}}{\sigma} \tag{8.66}$$

where $\overline{C}$ is the mean of C_k over all k, and σ the corresponding standard deviation.

The second measure is *betweenness centrality* (Z. HU et al., 2007) which for a network N is given by

$$B_i = \sum_{j,k \in N, j \neq k} \frac{n_{jk(i)}}{n_{jk}} \tag{8.67}$$

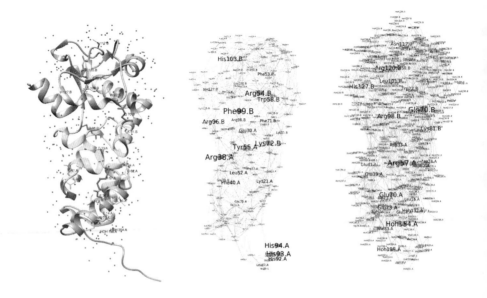

Figure 8.8 (**See color insert.**) Structure and RINs of the colicin E2 DNase-Im2 complex. For all three images, the residues at the interface between the two chains are colored in blue, and central residues are colored in a gradient from yellow (Z-score ≥ 2) to red (Z-score ≥ 4). For the explanation, see the text. (From G. BRYSBAERT et al., 2018.)

where n_{jk} is the number of shortest paths connecting residues j and k, and $n_{jk}(i)$ is the number of shortest paths between j and k that pass through residue i.

Figure 8.8 displays residue interaction networks for an enzymatic complex; relevant residues are selected by a Z-score (score value - mean/standard deviation). In this figure, water molecules surrounding the protein are considered as part of the network as the presence of specific water molecules can play a role in the function of the protein. The left figure shows a cartoon representation of the E2/Im2 complex (PDB-ID 3U43), with chain B (E2) located at the top and chain A (Im2) at the bottom; E2 is depicted in a darker shade while Im2 is lighter. Water molecules are shown as small red spheres. Central residues are drawn in stick representation; the depicted central residues are the additional ones that are highlighted in the right-hand network. The middle figure shows the residue interaction network generated from the structure, excluding water molecules. Finally, the right graph displays the differential residue interaction network of the 'wet' vs. 'dry' network, highlighting only the additional central residues.

Additional notes

Intrinsic vs. extrinsic noise. This topic enjoys growing interest in the bio-community. An application of this concept to the cell-fate decision system is by A. COLMAN-LERNER et al. (2005).

Parameter problems. A very general modeling problem in systems biology is the lack of knowledge of parameters; a problem of which already A. TUR-ING was well aware. A statistical mechanics approach to systems with many poorly known parameters was launched by K.S. BROWN and J.P. SETHNA (2003). SETHNA has continued this work in the direction of systems biology (GUTENKUNST et al., 2007).

Theory of pattern formation. *The* pioneer in the field is Alan Turing. A modern classic in the theory of pattern formation is the work by A. GIERER and H. MEINHARDT (1972).

The Min system. The Min system enjoys ongoing interest in modelling; see the papers by M. HOWARD, K. KRUSE, H. MEINHART, N. S. WINGREEN and collaborators. The system is nice since a lot can be done experimentally. Note, however, that despite the model approaches are all similar in the underlying philosophy, there are still unsolved questions, and the modeling is not unam-biguous. See the discussion by M. HOWARD and K. KRUSE (2005). Min is nice, but Drosophila is maybe nicer. Relating back to the problem of transcriptional control on the chromatin level it is of enormous interest and difficulty to study the structure formation in embryo development. Modern approaches employ modeling (J. REINITZ et al., 2003), microfluidics (E. M. LUCHETTA et al., 2005), and gene network engineering (M. ISALAN et al., 2005). (J. REINITZ has continued to work on this problem, it is worth to check out his more recent papers.

Statistical mechanics of networks. The literature on networks emerging within the statistical physics community has literally exploded since its in-ception. An early review was written by R. ALBERT AND A. L. BARABÁSI (2002). The mathematical theory of random graphs in the tradition of the pioneers (P. ERDÖS and A. RENYI, 1959), is summarized by B. BOLLOBÁS (2000). Two books cover this field, the first by M. NEWMAN (2018) (in second edition) and the second by A.-L. BARABÁSI (2016).

Protein-protein interaction maps. A high-quality protein 'interactome' map for yeast was published by Y. HU et al. (2008). The possible role of interactome maps for the treatment of humane disease was highlighted by M. VIDAL et al. (2011). Combining network analyses and phenotypes allows to turn the undirected into a directed network (A. VINAYAGAM et al., 2014). Conclusions that can be drawn from interactome maps for the evolution of genomes are reviewed in M.A. GHADIE et al. (2018). The usefulness of the approach for the detection of drug-disease relationships is discussed in F. CHENG et al. (2018).

Residue interaction networks. A detailed protocol for the use of network analyzing software is by DONCHEVA et al. (2012). A recent review on the topic is (L. DI PAOLA et al., 2013). A method combining network analyses with machine learning is by M. GIOLLO et al. (2014).

Network motifs. The existence of motifs can be used to construct an algorithmic procedure for the detection and significance decision of local network elements. The deterministic motifs are generalized to probabilistic motifs in which edges occur with a certain likelihood. This procedure, called *graph alignment* due to its analogy to sequence alignment, can be based on a suitably defined *scoring function*, i.e., a Hamiltonian for the motifs, H_{score} (J. BERG AND M. LÄSSIG, 2004). A recent survey of the field of graph alignment is by A. ELMSALLATI et al. (2016).

Finally, it should be stressed that the overrepresentation of certain motifs does not at all guarantee that these are the important building blocks in the system. Biological networks, after all, have evolved over the course of many thousands of years, and the networks hence carry a history along which is very difficult to take into account the present description of networks (see, e.g., E. DEKEL et al. (2005) on the selection of the feedforward motif). As useful as such notions as motifs and modules are as a work guide at present (for a discussion, see the paper by L. H. HARTWELL et al. (1999), as little is known about their true biological significance.

References

R. Albert, H. Jeong, A.-L. Barabási
Error and attack tolerance of complex networks
Nature **406**, 378-380 (2000)

R. Albert, A.-L. Barabási
Statistical mechanics of complex networks
Rev. Mod. Phys. **74**, 47-97 (2002)

U. Alon
Introduction to Systems Biology and the Design Principles of Biological Networks
Chapman & Hall/CRC (2006)

A.-L. Barabási, R. Albert
Emergence of scaling in random networks
Science **286**, 509-512 (1999)

A.-L. Barabási
Network Science
Cambridge University Press (2016)

J. Berg, M. Lässig
Correlated random networks
Phys. Rev. Lett. **89**, 228701 (2002)

J. Berg, M. Lässig
Local graph alignment and motif search in biological networks
Proc. Natl. Acad. Sci. USA **101**, 14689-14694 (2004)

B. Bollobás
Random Graphs
Academic Press (2000)

K.S. Brown, J. P. Sethna
Statistical mechanical approaches to models with many poorly known parameters
Phys. Rev. E **68**, 021904 (2003)

G. Brysbaert, R. Blossey, M.F. Lensink
The inclusion of water molecules in residue interaction networks identifies additional central residues
Frontiers Mol. Bio. **5**, #88 (2018)

F. Cheng et al.
Network-based approach to prediction and population-based validation of in silico drug repurposing
Nat. Comm. **9**, 2691 (2018)

A. Colman-Lerner, A. Gordon, E. Serra, T. Chin, O. Resnekov, D. Endy, C.G. Pesce, R. Brent
Regulated cell-to-cell variation in a cell-fate decision system
Nature **437**, 699-706 (2005)

E. Dekel, S. Mangan, U. Alon
Environmental selection of the feed-forward loop circuit in gene-regulation networks
Phys. Biol. **2**, 81-88 (2005)

L. Di Paoloa, M. De Ruvo, P. Paci, D. Santoni, A. Giuliani
Protein contact networks: an emerging paradigm in chemistry
Chem. Rev. **113**, 1598-1613 (2013)

N. T. Doncheva, Y. Assenov, F.S. Domingues, M. Albrecht
Topological analysis and interactive visualization of biological networks and protein structures
Nat. Protocols **7**, 670-685 (2012)

J.L. England, J. Cardy
Morphogen gradient from a noisy source
Phys. Rev. Lett. **94**, 078101 (2005)

A. Elmsallati, C.N. Clark, J. Kalita
Global Alignment of Protein-Protein Interaction Networks: A Survey
IEEE-ACM Transactions on Computational Biology and Bioinformatics **13**, 689-705 (2016)

M. B. Elowitz, A.J. Levine, E.D. Siggia, P.S. Swain
Stochastic gene expression in a single cell
Science **297**, 1183-1186 (2002)

P. Erdös, A. Renyi
On random graphs
Publ. Math. Debrecen **6**, 290-297 (1959)

A. Gierer, H. Meinhardt
A theory of biological pattern formation
Kybernetik **12**, 30-39 (1972)

M. Giollo, A.J.M. Martin, I. Walsh, C. Ferrari, S.C.E. Tosatto
NeEMO: a method using residue interaction networks to improve prediction of protein stability upon mutation
BMC Genomics **15** (Suppl 4): S7 (2014)

M.A. Ghadie, J. Coulombe-Huntington, Y. Xia
Interactome evolution: insights from genome-wide analyses of protein-protein interactions
Curr. Op. Struct. Biol. **50**, 42-48 (2018)

J.B. Gurdon, P.Y. Bourillot
Morphogen gradient interpretation
Nature **413**, 797-803 (2001)

R.N. Gutenkunst et al.
Universally sloppy parameter sensitivities in systems biology models
PLoS Comp. Bio. **3**, 1871-1878 (2007)

J.-D. Han, D. Dupuy, N. Bertin, M.E. Cusick, M. Vidal
Effect of sampling on topology predictions of protein-protein interaction networks
Nat. Biotech. **23**, 839-844 (2005)

L.H. Hartwell, J.J. Hopfield, S. Leibler, A.W. Murray
From molecular to modular cell biology
Nature **402**, C47-C50 (1999)

B. Houchmandzadeh, E. Wieschaus, S. Leibler
Establishment of developmental precision and proportions in the early Drosophila embryo
Nature **415**, 798-802 (2002)

M. Howard, A.D. Rutenberg, S. de Vet
Dynamic compartmentalization of bacteria: accurate cell division in E. Coli
Phys. Rev. Lett. **87**, 278102 (2001)

M. Howard, K. Kruse
Cellular organization by self-organization: mechanisms and models for Min protein dynamics
JCB **168**, 533-536 (2005)

M. Howard
A mechanism for polar protein localization in bacteria
J. Mol. Biol. **335**, 655-663 (2004)

M. Howard, P. Rein ten Wolde
Finding the center reliably: robust patterns of developmental gene expression
Phys. Rev. Lett. **95**, 208103 (2005)

Y. Hu et al.
High-quality binary protein interaction map of the yeast interactome network
Science **322**, 104-110 (2008)

Z. Hu, D. Bowen, W. Southerland, A. del Sol, Y. Pan, R. Nussinov, B. Ma
Ligand binding and circular permutation modify residue interaction network in DHFR
PLoS Comp. Bio. **3**, e117 (2007)

K. Huang, Y. Meir, N.S. Wingreen
Dynamic structures in Escherichia coli: spontaneous formation of MinE rings and MinD polar zones
Proc. Natl. Acad. Sci. USA **100**. 12724-12728 (2003)

M. Isalan, C. Lermerle, L. Serrano
Engineering gene networks to emulate drosophila embryonic pattern formation
PLoS Biology **3**, e64 (2005)

H. Jeong, B. Tombor, R. Albert, Z.N. Oltvai, A.-L. Barabási
The large-scale structure of metabolic networks
Nature **407**, 651-654 (2000)

H. Jeong, S.P. Mason, A.-L. Barabási, Z.N. Oltvai
Lethality and centrality in protein networks
Nature **411**, 41-42 (2001)

S. Kalir, S. Mangan, U. Alon
A coherent feed-forward loop with a SUM input function prolongs flagella expression in Escherichia coli
Mol. Syst. Biol. msb44100010 (2005)

J. Kim, P.L. Krapivsky, B. Kahng, S. Redner
Infinite-order percolation and giant fluctuations in a protein interaction network
Phys. Rev. E **66**, 055101 (2002)

M. Loose, E. Fischer-Friedrich, C. Herold, K. Kruse, P. Schwille
Min protein patterns emerge from rapid rebinding and membrane interaction of MinE
Nat. Struct. Mol. Biol. **18**, 577-584 (2011)

E.M. Luchetta, J.H. Lee, L.A. Fu, N.H. Patel, R.F. Ismagilov
Dynamics of Drosophila embryonic patterning network perturbed in space and time using microfluidics
Nature **434**, 1134-1138 (2005)

S. Mangan, U. Alon
Structure and function of the feed-forward loop network motif
Proc. Natl. Acad. Sci. USA **100**, 11980-11985 (2003)

S. Maslov, K. Sneppen
Specificity and stability in topology of protein networks
Science **296**, 910-913 (2002)

G. Meacci, K. Kruse
Min-oscillations in Escherichia coli induced by interactions of membrane-bound proteins
Phys. Biol. **2**, 89-97 (2005)

H. Meinhardt, P.A.J. de Boer
Pattern formation in Escherichia coli: a model for the pole-to-pole oscillations of Min proteins and the localization of the division site
Proc. Natl. Acad. Sci. USA **98**, 14202-14207 (2001)

R. Milo, S.S. Shen-Orr, S. Itzkovitz, N. Kashtan, D. Chklovskii, U. Alon
Network motifs: simple building blocks of complex networks
Science **298**, 824-827 (2002)

M.E.J. Newman
Random graphs as models for networks
in *Handbook of Graphs and Networks*, S. Bornholdt and H. G. Schuster (eds.), (2003)

M.E.J. Newman, C. Moore, D.J. Watts
Mean-field solution of the small-world network model
Phys. Rev. Lett. **84**, 3201-3204 (2000)

M. Newman
Networks: An Introduction
Oxford University Press, 2nd ed. (2018)

J. Park, M.E.J. Newman
The statistical mechanics of networks
Phys. Rev. E **70**, 055101 (2004)

E. Ravasz, A.L. Somera, D.A. Mongru, Z.N. Oltvai, A.-L. Barabási
Hierarchical organization of modularity in metabolic networks
Science **297**, 1551-1555 (2002)

J. Reinitz, S. Hou, D.H. Sharp
Transcriptional control in Drosophila
ComPlexUs **1**, 54-64 (2003)

S. Shen-Orr, R. Milo, S. Mangan, U. Alon
Network motifs in the transcriptional network of Escherichia coli
Nat. Genetics **31**, 64-68 (2002)

A. del Sol, H. Fujihashi, D. Amoros, R. Nussinov
Residue centrality, functionally important residues, and active site shape: analysis of enzyme and non-enzyme families
Prot. Sci. **15**, 2120-2128 (2006)

D. Sprinzak, M. B. Elowitz
Reconstruction of genetic circuits
Nature **438**, 443-448 (2005)

P.S. Swain, M.B. Elowitz, E.D. Siggia
Intrinsic and extrinsic contributions to stochasticity in gene expression
Proc. Natl. Acad. Sci. USA **99**, 12795-12800 (2002)

M. Vidal, M.E. Cusick, A.-L. Barabási
Interactome networks and human disease
Cell **144**, 986-998 (2011)

A. Vinayagam et al.
Integrating protein-protein interaction networks with phenotypes reveals signs of interactions
Nat Methods **11**, 94-99 (2014)

A. Wagner
The yeast protein interaction network evolves rapidly and contains few redundant duplicate genes
Mol. Biol. Evol. **18**, 1283-1292 (2001)

D.J. Watts, S.H. Strogatz
Collective dynamics of 'small-world' networks
Nature **393**, 440-442 (1998)

D. M. Wolf, A. P. Arkin
Motifs, modules and games in bacteria
Curr. Op. Microbiol. **6**, 125-134 (2003)

E. Yeger-Lotem, S. Sattath, N. Kashtan, S. Itzkovitz, R. Milo, R.Y. Pinter,
U. Alon, H. Margalit
Network motifs in integrated cellular networks of transcription-regulation and protein-protein interaction
Proc. Natl. Acad. Sci. USA **101**, 5934-5939 (2004)

Index